T0280827

Cambridge Elements ≡

Elements in the Philosophy of Mathematics
edited by
Penelope Rush
University of Tasmania
Stewart Shapiro
The Ohio State University

PHENOMENOLOGY AND MATHEMATICS

Michael Roubach
Hebrew University of Jerusalem

CAMBRIDGE
UNIVERSITY PRESS

Shaftesbury Road, Cambridge CB2 8EA, United Kingdom

One Liberty Plaza, 20th Floor, New York, NY 10006, USA

477 Williamstown Road, Port Melbourne, VIC 3207, Australia

314–321, 3rd Floor, Plot 3, Splendor Forum, Jasola District Centre,
New Delhi – 110025, India

103 Penang Road, #05–06/07, Visioncrest Commercial, Singapore 238467

Cambridge University Press is part of Cambridge University Press & Assessment,
a department of the University of Cambridge.

We share the University's mission to contribute to society through the pursuit of
education, learning and research at the highest international levels of excellence.

www.cambridge.org
Information on this title: www.cambridge.org/9781009462501

DOI: 10.1017/9781108993913

First published 2023

A catalogue record for this publication is available from the British Library

ISBN 978-1-009-46250-1 Hardback
ISBN 978-1-108-99539-9 Paperback
ISSN 2399-2883 (online)
ISSN 2514-3808 (print)

Cambridge University Press & Assessment has no responsibility for the persistence
or accuracy of URLs for external or third-party internet websites referred to in this
publication and does not guarantee that any content on such websites is, or will
remain, accurate or appropriate.

Phenomenology and Mathematics

Elements in the Philosophy of Mathematics

DOI: 10.1017/9781108993913
First published online: November 2023

Michael Roubach
Hebrew University of Jerusalem

Author for correspondence: Michael Roubach, roubach@mail.huji.ac.il

Abstract: This Element explores the relationship between phenomenology and mathematics. Its focus is the mathematical thought of Edmund Husserl, founder of phenomenology, but other phenomenologists and phenomenologically oriented mathematicians, including Weyl, Becker, Gödel, and Rota, are also discussed. After outlining the basic notions of Husserl's phenomenology, the author traces Husserl's journey from his early mathematical studies to his mature phenomenology. Phenomenology's core concepts, such as intention and intuition, each contributed to the emergence of a phenomenological approach to mathematics. This Element examines the phenomenological conceptions of natural number, the continuum, geometry, formal systems, and the applicability of mathematics. It also situates the phenomenological approach in relation to other schools in the philosophy of mathematics – logicism, formalism, intuitionism, Platonism, the French epistemological school, and the philosophy of mathematical practice.

Keywords: phenomenology, mathematics, intuition, constitution, Husserl

ISBNs: 9781009462501 (HB), 9781108995399 (PB), 9781108993913 (OC)
ISSNs: 2399-2883 (online), 2514-3808 (print)

Contents

1 Basic Concepts of Husserl's Phenomenology

Phenomenology is one of the major schools in twentieth-century philosophy.[1] It can be said to originate in Husserl's *Logical Investigations* (*LI*), written in 1900–1. Phenomenology focuses on how things are given to consciousness, with special emphasis on their structure. It emerged as part of Husserl's response to the challenge posed by the psychologistic construal of logic and mathematics. Psychologism, as Husserl understands it, maintains that the meanings of logical notions, such as *syllogism*, and logical connectives, such as conjunctions, are given through study of the mind (*LI*, vol. 1, §17). For example, psychologism takes the justification of the law of noncontradiction to be the human mind's inability to think a proposition and its contrary together (§25). According to Husserl, meeting the challenge of establishing the validity of the basic laws of logic nonpsychologistically – that is, without grounding that validity in contingent facts about the human mind – requires a clear distinction between the subjective and the objective, for example, between acts of judging and the ideal content of propositions (§47). Husserl further contends that separating ideal content from its apprehension by the mind does not suffice to overcome psychologism. If such a separation entails that the grasping of ideal content is an act of a contingent human mind, then it would seem that, ultimately, knowledge of the meanings of logical notions is dependent on contingencies, undermining the ideality of these meanings. How, then, Husserl asks, "can the ideality of the universal *qua* concept or law enter the flux of real mental states and become an epistemic possession of the thinking person?" (*LI*, Introduction to vol. 2 of the German edition, §2). Husserl therefore argues that a nonpsychologistic account of logic and mathematics requires a new epistemology, which is provided by phenomenology.

The two cornerstones of phenomenology are intentionality and intuition. Intentionality is directedness toward an object: intention is always about something. An intentional act has three distinct, but connected, parts: the act, its content (the object as it is intended), and the object toward which the act is directed. For example, the intentional act of thinking that a certain table is round is an act of thought; its content is the proposition that the table is round; the object toward which it is directed, that is, the object it is about, is the table. To understand how this notion enables us to meet the challenge of providing a nonpsychologistic basis for logic, a comparison with representation will be helpful. Both representation and intention involve a relation between

---

[1] This Element will not address earlier versions of phenomenology, such as Hegel's, or issues pertaining to phenomenology that are connected to Husserl only loosely, if at all (e.g., use of the term "phenomenology" in the analytic tradition).

consciousness, content, and an object, but in the case of intention, as Husserl understands it, the content is not private and specific to an individual consciousness, since through that content some object is given to a consciousness as being of a certain kind. For example, in order to perceive something, it must be given as perceptible, and therefore the sense (*Sinn*) of perceptibility is part of the content of the intention directed toward perceiving the thing. In Husserl's words, all perceptions have something in common; he calls this something their quality (*LI* V, §20). Moreover, in an intentional act, the consciousness in question is not only linked to content that is not a subjective representation, but, given that the intentional act as a whole is directed toward an object, also refers to an object.

Husserl's notion of intentionality and the relation between an intentional act and its object calls to mind Frege's distinction between meaning or sense, and reference. But whereas Husserl takes the relation between meaning and reference (the aboutness of a term or expression) to require a consciousness directed at the object referred to, for Frege, an expression's aboutness is determined by the meanings of its component terms; reference does not require consciousness.

The second cornerstone of phenomenology is intuition. Intuition is the ultimate justification of knowledge: "every originary presentive intuition is a legitimizing source of cognition" (*Ideas*, §24). Husserl calls this role of intuition "the principle of all principles." He contrasts intuition to "merely symbolic understanding" (*LI*, Introduction to vol. 2 of the German edition, §2). For Husserl, intuitions are the fulfillments of intentional acts. Intuitions function as confirmation or evidence for the meanings that constitute the content of intentional acts. In intuition, there is also an element of recognizing something as something. An intuition of a chair involves recognizing it as a chair; this recognition confirms the meaning of "there is a chair in the room." As such, Husserlian intuitions are not just sensory, but involve meanings. In the case of intuiting a chair, the meaning in question is what it is to be a chair, the essence of chairness. In *Logical Investigations*, nonsensory intuition is called "categorial intuition" (*LI* VI, §§40–58). Categorial intuition grasps not only empirical essences such as "chairness" but also logical categories such as "and" and "is." For example, an intuition that fulfills the proposition "the gold is yellow" involves intuiting the copula, inasmuch as what is intuited is not only a piece of gold and yellowness, but there is also an intuition of the gold's being yellow (§44). Intuition of an aggregate involves intuition of "and." We will return to the idea of categorial intuition in Section 3.2, since it plays a pivotal role in phenomenological approaches to mathematics.

Intuition is pertinent to one of the central maxims of phenomenology: "we must go back to the 'things themselves'" (*LI*, Introduction to vol. 2 of the

German edition, §2).[2] It is through intuition that things are given as they are, and through intuition that we experience the world as it is. There is no veil between us and the world. Husserl's phenomenology thus rejects Kant's distinction between things as we experience them and things as they are in themselves.

In some respects, Husserl's phenomenology has an affinity with empiricism in general and Hume's position in particular. A key tenet of phenomenology is that in examining our experience of the world, we are not to make any metaphysical assumptions, a principle that is consonant with empiricism. Husserl also adopts Hume's distinction between relations of ideas and matters of fact. But on the other hand, Husserl criticizes the empiricist attempt to completely eliminate the assumption of any general concepts. General concepts cannot be arrived at by a process of abstraction from individual objects, since in such a process of abstraction, the common feature is already assumed and recognized (*LI* II, §§4, 29). Rather, general concepts are apprehended directly, by way of intentionality and intuition. The direct apprehension of general concepts is crucial for the repudiation of psychologism, since it means that the claim that a consciousness can grasp logical and mathematical concepts does not rule out ascribing ideality to those concepts. To put it in Humean terms, Husserl maintains that relations of ideas can be grasped as such.

Husserl's exposition of phenomenology in *LI* remained unclear on two important points: the ontological status of ideal objects, and phenomenology's relation to psychology. As to the former, Husserl does not argue that the direct apprehension of ideas implies their metaphysical hypostasis (*LI* I, §31). Phenomenology makes no ontological commitment to the independent existence of universals outside of thought (*LI* II, §7). But ideal objects do "exist genuinely" (*LI* II, §8). Husserl's stance on the ideality of general concepts in *Logical Investigations* was often interpreted as Platonistic, an interpretation he explicitly rejected (e.g., in Husserl 1975, §4). The second point regarding which the exposition in *Logical Investigations* was deemed insufficiently clear is Husserl's stance on psychology. On the one hand, Husserl criticizes psychologism, and rejects empirical psychology as a foundation for logic, but on the other, phenomenology is characterized as descriptive psychology, and the distinction between descriptive and empirical psychology is far from sharp.

These two ambiguous points played a key role in bringing about an important change in the method of phenomenological inquiry, a change widely referred to as "the transcendental turn." On the revised conception, inquiry into the mode of things' givenness starts from what Husserl calls "the natural attitude," the attitude one ordinarily takes toward the world. This noncritical attitude assumes

---

[2] The German word Husserl uses for "things" in this phrase is "*Sachen*," not "*Dinge*."

that the things in the world exist. Phenomenological inquiry begins by suspending this assumption. Husserl, borrowing a term from ancient skepticism, calls this suspension "*epoché*" (*Ideas*, §32). Another name for this suspension is "phenomenological reduction," since it shifts the focus from things to the way things are given to consciousness, for example, from a perceived table to the table as it is perceived.

Phenomenological reduction provides a solution to the two ambiguities. It allows for a clear distinction between phenomenology and psychology. Psychology is a positive science that presupposes the existence of the objects it investigates; phenomenology is not. It follows that phenomenological reduction (bracketing) also obviates the claim that phenomenology is Platonistic, since, given that phenomenology does not presuppose the existence of any objects it investigates, it does not presuppose the existence of ideal objects.

The transcendental turn in phenomenology is accompanied by the introduction of a distinction between noesis and noema. Noesis refers to the subjective side of an intention. More precisely, it refers to apprehension of the intention's content. Noema refers to the intention's content, that is, to the intentional object as intended. According to Husserl, there is a correlation between noesis and noema (*Ideas*, §88). Each act of apprehension (noesis) is related to a content (noema), and no content is separate from its apprehension. The inseparability of noesis and noema is a further aspect of phenomenology's non-Platonism.

Another important step in the phenomenological inquiry is the extraction of essences through a process Husserl calls "eidetic reduction" (*Britannica*, §4). Eidetic reduction uses imagination to create permutations of a given experience and extract the component common to all of them. For example, we can change an experience of a table by turning the table around in our imagination and then extracting the experience's essential elements. The experience of an object in space is always given only partially, from one point of view. In any act of perceiving a table, there are sides of the table that remain unseen, and this is the essence of the way spatial objects are given.

Through the process of eidetic reduction, phenomenology seeks to arrive at the essences underlying all our experiences, including scientific investigations. The essence of the experience of space, for instance, is relevant not only to the positive science of psychology, but also to other sciences in which the notion of space plays a role, such as physics. Although the notion of space in physics differs from the notion of space in psychology, they do share a core sense of spatiality. By extracting the essences that constitute the basic notions, phenomenology provides the basis for all sciences: "The creation of fundamental concepts is therefore, in the most literal sense, a *fundamental performance*, laying the foundations for all sciences" (*FTL*, §71; italics in the original).

In light of phenomenological reduction, the domain that is investigated by phenomenology is that of consciousness (*Ideas*, §33). Consciousness is characterized primarily in terms of intentionality. A crucial activity of consciousness is to constitute the senses of the content given to it (*Ideas*, §86).

With respect to constitution, Husserl distinguishes between static and genetic phenomenology. Static phenomenology focuses on the description of constituted essences. Genetic phenomenology attends to the origins and development of the constitution of essences, which must be distinguished from the *causal* generation of essences in empirical consciousness. The notion of time plays an important role in the genesis of essences, as does the notion of association (*Britannica*, §5).

The last phase of Husserl's phenomenology addresses the tension between modern mathematical physics, in which there is no place for the subjective side of experience (e.g., the experience of secondary qualities) and the life-world, the world as we experience it in everyday life. In *The Crisis of European Sciences*, Husserl describes how this tension created a crisis with regard to the foundations of science: which of the two ultimately anchors the validity of the sciences? Arguing that only our experience of the world can serve as such an anchor, Husserl's later works ground all knowledge in the life-world.

2 Husserl's Path from Mathematics to Phenomenology

Husserl studied mathematics in Leipzig and Berlin, and after writing a thesis on the calculus of variation and receiving his PhD from the University of Vienna, served as an assistant to Weierstrass, one of the era's leading mathematicians.[3] His interest in the philosophical foundations of mathematics led him to study with Brentano, whose approach was psychological, inasmuch as it examined philosophy from the perspective of the human mind and its capacities. Husserl's studies with Brentano, and subsequent work with Brentano's student Carl Stumpf in Halle, culminated in Husserl's 1887 habilitation thesis on the concept of natural number, which was incorporated into his first book, *The Philosophy of Arithmetic* (*PA*), published in 1891.

Husserl takes the primary feature of the natural numbers to be cardinality (*PA*, 12) and, following Weierstrass, takes the natural numbers to be the basis for all the other kinds of numbers (whole, rational, real, etc.). His research on the natural numbers aims "to satisfy not merely the arithmetical interests, but the logical and psychological interests above all" (14). Natural numbers are characterized in terms of multiplicities, as "collective combinations" of "one"s (78, 81).

---

[3] This section covers Husserl's thought on mathematics up to *Logical Investigations*, his phenomenological breakthrough.

The "one"s are arrived at through a process of abstracting from any object, from any something. There are no restrictions on what can be counted: any object that can be thought, including imaginary objects, can be counted. Nor is there a criterion for the unification of these objects. There need not be (as Frege, for example, argued) a feature common to the collection: "collective unification is not intuitively given *in* the representation content, but instead has its subsistence only in certain psychical acts that embrace the contents in a unifying manner" (77; italics in the original). The unity is the outcome of an act, hence the collective combination can be grasped only by reflecting on the specific act in which the multiplicity is unified (77). For Husserl, this sort of combination, which is not based on a common feature, is conveyed by the word "and," which, he points out, can also connect terms (78). To arrive at the concept of "one" (*Eins*), it is necessary to abstract from the specific content of each thing in a collection of things, and take them as just plain somethings (*Etwas*); reflection on these undifferentiated somethings generates the concept of "one." The concept of any natural number is a combination of such "one"s. For example, the concept of the number 2 is "one and one," the concept of the number 3 is "one and one and one," and so on (85).

Another feature of Husserl's approach to the natural numbers in *Philosophy of Arithmetic* that, like cardinality, remains important in his later work is the salience of the nexus between seeking the origin of a notion and relating this origin to the notion's givenness in intuition. Husserl's account of the natural numbers starts out from the notion of authentic numbers, which can be given in intuition, and then proceeds to symbolic numbers. Privileging the authentic and seeking ways to ground the symbolic in intuition will remain a central aspect of Husserl's phenomenology, though not in the form it takes in *Philosophy of Arithmetic*.[4]

But there are also differences between Husserl's views as set out in *Philosophy of Arithmetic* and his subsequent thinking. These differences pertain both to Husserl's conception of mathematics and to his general philosophical approach to matters of logic. With regard to mathematics, even while he was writing *Philosophy of Arithmetic*, Husserl was concerned about the problem of how to understand other kinds of numbers (negative whole numbers, fractions) as extensions of the natural numbers. This problem, to which mathematicians in the second half of the nineteenth century gave much attention, is crucial for the meaning of the notion of number itself, and for the meaning of the arithmetical operations. For example, does subtraction have the same meaning vis-à-vis the

---

[4] On the role of the distinction between authentic and symbolic presentations of numbers in Husserl's post-*Philosophy of Arithmetic* account of numbers, see Miller (1982).

natural numbers and the whole numbers? Husserl discusses this in an 1891 letter to his supervisor Stumpf: "If it [arithmetic] deals with discrete magnitudes, then 'fractions,' 'irrational numbers,' imaginary numbers and, in the case of the cardinal numbers for example, the negative numbers also lose all sense" (Husserl 1994, 15). The problem is especially acute given Husserl's account of natural number, which interprets all natural numbers (including those given only symbolically) as multiplicities of units. The source of the problem is the connection between the natural numbers' authentic and symbolic givenness, which imposes a specific interpretation of the natural numbers – namely, that they are multiplicities of units – on all the other kinds of numbers, though fractions, for example, cannot be conceived as multiplicities of units.

Given this problem, Husserl's position on the connection between authentic and symbolic givenness changed. There is already evidence of the change in the last part of *Philosophy of Arithmetic*, where Husserl separates the symbolic levels of numbers (numerals) from any specific concept of number (*PA*, 253).

Husserl was also preoccupied with a related problem concerning the extension of the natural numbers to other systems of numbers: expressions that have no meaning in one system of numbers but are meaningful in extensions of that system. For instance, the expression "5–7" has no meaning in the natural numbers, but does have meaning in the system of the whole numbers, which can be interpreted as an extension of the natural numbers. Hence, we cannot speak literally of extending the natural numbers to the whole numbers – to give just one example – but the natural numbers and the whole numbers are nonetheless related.

Husserl's solution to this problem, presented in two lectures to the Mathematical Society of Göttingen (often called "the Double Lecture") in 1901, is to introduce a means of connecting two domains of numbers (e.g., the natural numbers and the whole numbers), provided two conditions are satisfied. The first condition is that the larger domain has to be consistent; the second condition is that "every proposition falling within that [i.e., the original] domain is either true on the basis of the axioms of that domain or else is false on the same basis (i.e., is contradictory to the axioms)" (*PA*, 428).[5] Satisfaction of these conditions ensures that any proposition in the new domain that is equivalent to a proposition in the original domain will have the same truth value as the original proposition. For example, if the proposition $2 + 2 = 4$ is true (or provable) in the natural numbers, then its equivalent in the whole numbers is also true (or provable). This solution does not require that the signs in the two

---

[5] Husserl, like the other philosophers of this period, does not distinguish between the syntactic and the semantic aspects of a formal system. Proof of a proposition justifies accepting that proposition as true.

systems be given the same meanings: the natural number 2 need not have the same meaning as the whole number 2. Husserl has thus provided a solution to the problem of extending one system of numbers to another.

Husserl further distinguishes between relative and absolute definiteness, defining the former as follows: "An axiom system is relatively definite if every proposition meaningful according to it is decided under restriction to its domain" (427). The definiteness is relative, since extension of the domain is possible. He distinguishes this from absolute definiteness: a system is absolutely definite "if every proposition meaningful according to it is decided in general" (427). A system has absolute definitude if it is uniquely defined by its axioms (426). Husserl compares this notion to Hilbert's characterization of an axiomatic system's completeness. According to Hilbert, a system of axioms is complete if it is "incapable of being extended while continuing to satisfy all the axioms" (Hilbert 1996, 1094).[6]

The change in Husserl's thinking regarding the idea that all of mathematics could be based on "authentic" natural numbers is related to his development of the idea of a manifold. A manifold is a multiplicity of objects each of which is determined by the same set of laws. The objects are not determined individually, but only in relation to other objects. Riemann's theory of geometric manifolds was the inspiration for this concept of the manifold, but Husserl applies it to all the domains of mathematics, including the natural numbers (*LI*, Prolegomena, §70). In *Formal and Transcendental Logic*, Husserl defines a manifold or multiplicity thus: "Multiplicity meant properly the *form-idea of an infinite object-province for which there exists the unity of a theoretical explanation* or, in other words, the unity of a *nomological science*" (*FTL*, §31; italics in the original). The different kinds of numbers (natural, whole, imaginary, etc.) are separate manifolds, and the nonnatural numbers do not develop out of the natural numbers. With this theoretical shift, Husserl has, to borrow a Hilbertian distinction (Hilbert 1996, 1092–3), moved from a genetic account of mathematics to an axiomatic one.

Although each manifold is self-standing, it can be connected to other manifolds through the concept of definiteness. Manifolds are not relevant just to mathematics, but can be adduced to explicate all formal notions, including logical notions. For example, in *Logical Investigations*, Husserl develops a formal theory of part–whole relations that, although Husserl does not say this explicitly, can itself be interpreted

---

[6] There is a vast literature on Husserl's notion of definiteness. For background on Husserl's development of the notion of definiteness, see Hartimo (2007). For a reconstruction of Husserl's position in his Göttingen "Double Lecture," see Centrone (2010, chapter 3). On the relation between Husserl's notion of definiteness and Hilbert's notion of completeness, see Majer (1997) and da Silva (2000, 2016).

as a kind of manifold. Husserl calls the domain of manifolds as a whole the *mathesis universalis*,[7] which he characterizes as "including all *a priori*, categorial knowledge in the form of systematic theories" (*LI*, Introduction to vol. 2 of the German edition, §7; italics in the original). That is, the *mathesis universalis* is equated with all the theories that are fully deductive, theories whose domains are definite manifolds (*FTL*, §35b). The *mathesis universalis* framework establishes the unification of mathematics and logic, not in the logicist sense of reducing mathematics to logic, but in the sense of showing that formal mathematics and logic belong to the same family. Both fields are formal in the sense that they apply to any object, as their objects are "somethings in general" (*Etwas überhaupt*).

Interpreting the natural numbers as one type of manifold within the larger framework of the *mathesis universalis* is the first of two major changes in Husserl's post-*Philosophy of Arithmetic* account of number. The second major change in Husserl's position concerns the relation between logic and psychology, a relation that was not articulated in *Philosophy of Arithmetic*, but became central in *Logical Investigations*.[8] As Husserl acknowledges (e.g., in *PA*, 22), the account of number in *Philosophy of Arithmetic* is psychological inasmuch as it seeks the origin of our concept of natural number. As he later observes, it can also be seen as a phenomenological-constitutional account of the concept of natural number (*FTL*, §27a). We can say, though, that whereas from *Logical Investigations* onward, Husserl critiques psychologism and distinguishes phenomenology from empirical psychology, in *Philosophy of Arithmetic* he does not draw a clear distinction between psychology and logic. Indeed, this failing was the main target of Frege's critique in his 1894 review of *Philosophy of Arithmetic*: Frege claims that the book does not make clear the distinction between the subjective and the objective (Frege 1972, 324–5).[9]

---

[7] The phrase "*mathesis universalis*" originates in Barocius' Latin translation of Proclus. A combination of Greek and Latin, its literal meaning is "universal learning." The term, which became well known through its invocation by Descartes in *Rules for the Direction of the Mind*, and later, by Leibniz in *New Essays on Human Understanding*, has been understood in different ways. Descartes used it to designate the features common to the various fields of mathematics, especially geometry and arithmetic. Leibniz used it to designate the formal disciplines in general. Husserl links his own notion of the *mathesis universalis* to Leibniz's (*LI*, Prolegomena, §60).

[8] Husserl mentions this change in his Foreword to the first edition of *Logical Investigations* (*LI*, vol. 1, 2).

[9] The question of the role of Frege's critique in the evolution of Husserl's position was the subject of lively discussion in the second half of the twentieth century. As Mohanty (1977) showed, there is evidence of a change in Husserl's position regarding the distinction between logic and psychology even before Frege's review. Willard, too, contends that the change in Husserl's position was not a response to Frege's critique, arguing that it arose from Husserl's dissatisfaction with the distinction between authentic and symbolic presentation (Willard 1980, 61). Nevertheless, according to Boyce Gibson, Husserl told him that he thought Frege's critique "hit the nail on the head" (Spiegelberg 1971, 66). Hence, even if Frege's critique did not prompt the change in Husserl's position, it served to affirm it.

The shift in Husserl's views on the logic–psychology nexus has several aspects. First, Husserl distinguishes between meaning and presentation. Meaning is ideal, transcending the subjective realm, whereas presentations are real entities that are part of the consciousness that experiences them. The meaning of an assertion, for instance, is ideal, and must be distinguished from the various acts of judging that what is asserted is the case (*LI*, Introduction to vol. 2 of the German edition, §2). Second, in *Logical Investigations* Husserl distinguishes the object of an intention, which is not part of the intention, from the intentional object as intended, which is (*LI* V, §17). This distinction enables him to introduce the idea of intentions being fulfilled in – that is, made evident by – intuitions. Fulfillment provides evidence of the objective validity of an intention's content (the intentional object as intended). Intuitions thus provide evidence that renders logical and mathematical notions objectively valid. Third, Husserl's notion of intentionality involves a close link between consciousness of mathematical and logical concepts, and the content of those concepts. Although there is a clear distinction between an intentional act and its content, they are not separable. Such separation would risk reintroducing a psychological account of how these concepts are grasped. Hence Husserl's notion of intentionality is in a better position to meet the challenge of epistemically securing the validity of the basic laws of logic in a nonpsychologistic manner than is the stance that takes intentional content to be separable.[10]

What impact does Husserl's critique of psychologism have on the conception of the natural numbers set out in *Philosophy of Arithmetic*? As we saw, in *Logical Investigations*, the natural numbers are conceived as a manifold. But understanding numbers in terms of formal manifolds governed by axioms is not sufficient. In addition, numbers should also be understood as given in intuition as ideal species (*LI*, Prolegomena, §46). Specific numbers are objects, but not empirical objects. Rather, they are ideal singular species. For numbers are different from any specific group of things, and from any presentation in consciousness – for example, from any act of counting. The number five, say, can be distinguished both from any group of five things, and from any act of counting (§46). This characterization of the natural numbers reflects Husserl's conception of intention and its fulfillment in intuition. But how is the notion of an intention's "fulfillment" to be understood? Does it entail some sort of Platonic realm of mathematical objects? Or, alternatively, does it entail an Aristotelian view on which the intuited ideal species are part of the real world? These questions are discussed in the following sections.

---

[10] In this sense Husserl's phenomenology can be interpreted as providing an answer to Benacerraf's accessibility problem (Benacerraf 1973). Although logical and mathematical concepts are ideal and transcend the human mind, intentionality renders them accessible to cognition.

3 Phenomenology of Mathematics

To examine mathematics from the perspective of phenomenology is to take an approach that emphasizes issues related to cognition. The basic concepts of phenomenology – intention and intuition – are central to this examination.

The phenomenology of mathematics, or at least its Husserlian variant, moves between two poles. On the one hand, as we saw in the preceding sections, Husserl's theory of manifolds is indicative of a formal orientation toward mathematics. On the other hand, Husserl maintains that the formal in itself does not suffice to provide a full philosophical account of mathematics. There must also be a level that imparts meaning to mathematics' formal dimension. These two poles are related to Husserl's distinction between significative intention and intuitive intention (*LI* VI, §63). The statements of formal mathematics exemplify intentions that are purely significative, but nonetheless require a grounding in intuition.

We will begin our examination of the phenomenology of mathematics by looking at the relation between Husserl's theory of manifolds and phenomenology.

3.1 Husserl's *Mathesis Universalis* and Phenomenology

As we saw in Section 2, after *Philosophy of Arithmetic*, Husserl's position underwent two significant changes, one specifically connected to mathematics (the emergence of the theory of manifolds and abandonment of the genetic approach to mathematics) and the other more general (the critique of psychologism and development of phenomenology as an alternative to psychology). Are there links between these changes? Husserl himself thinks that there are: "My way to phenomenology was essentially determined by the *mathesis universalis*" (Husserl 1980, §10), but does not spell out the relations between them. In this section, I will suggest ways in which the notion of the *mathesis* is linked to the development of phenomenology. I will begin by examining the role of the theory of manifolds in the *Logical Investigations* framework.

The theory of manifolds is treated in Husserl's discussion of the idea of pure logic in the last chapter of the Prolegomena, before he proceeds to the six specific Investigations. The aim of the theory of manifolds is to work out "the form of the essential types of possible theories or fields of theory" (*LI*, Prolegomena, §70). The theory should, therefore, cover the theories of the basic concepts or categories of all scientific theories. Husserl divides these concepts or categories into two sorts: categories of meaning, such as "proposition," "truth," and so on, and categories of objects, such as "unity," "number," and so on. (*LI*, Prolegomena, §67). The basic constraint on these theories (discussed in *Formal and Transcendental Logic*) is that they must not contain any contradiction. However, they need not be true.

There are several links between the theory of manifolds and phenomenology. First, the structure of intentionality, with its clear distinction between intention and fulfillment, is reflected in the structure of manifolds. The applicability of any given manifold is distinct from the manifold as a formal structure, but, as in the case of the structure of intentionality, the relation to possible application is part of the logical sense of any formal mathematics (*FTL*, §40). Any given manifold refers to objects, but the determination that a specific set of objects is or is not an instance of the manifold is not part of the manifold itself. The objects of formal systems are "somethings in general," not specific objects. The links between intentionality and manifolds are likewise connected to the distinction between categories of meaning and categories of objects (Husserl also refers to the latter categories as constitutive of formal ontology), and the relation between the two sorts of categories. This distinction mirrors the structure of intentionality, within which the content of an intentional act is distinguished from the object at which the intentional act is directed. It also mirrors the distinction between the two dimensions of a theory: the linguistic dimension (what the theory says) and the ontological dimension (the objects the theory is about).

Second, I contend that the *mathesis universalis* plays a role in Husserl's adopting, from *Logical Investigations* onward, the idea of "the correlation between world ... and its subjective manners of givenness" (*Crisis*, §48). This correlation, which is exemplified in the fulfillment of an intention by an intuition, is reflected, in the *mathesis universalis*, in the correlation between formal ontology and apophantics, the science of articulated thoughts (judgments). Every axiomatic system has two dimensions: a linguistic dimension (the sentences of the axiomatic system) and an ontological dimension (the multiplicities that are the ontological counterparts of the system's sentences). In *The Crisis of European Sciences*, however, though Husserl continues to uphold the a priori correlation between thinking (modes of givenness) and being, he does not link it to the *mathesis universalis*.

Third, the links between phenomenology and the *mathesis universalis* are even deeper. They are in some sense complementary. If we can fully capture a basic notion such as natural number through a formal system, then phenomenology's role of providing epistemic access to that basic notion seems to lose its importance or even to be redundant. The need to intuit notions arises primarily in cases where they are not fully captured by formal theories.[11]

Fourth, the notion of definitude as characterizing an axiomatic system can also be connected to phenomenology. This point is related to the previous point.

---

[11] For a detailed discussion of the relation between the *mathesis universalis* and phenomenology, see Roubach (2022).

Hartimo (2007, 2018) argues that one of the aims of the notion of definiteness is to render axiomatic systems categorical.[12] Under this interpretation, the categoricity criterion, which means that the theory determines a notion up to isomorphism, must be met, since that ensures that a formal theory captures all the characterizations of a given notion. Hartimo (2018) points out that the notion of definitude is central to Husserl's critique of psychologism, as it provides the required framework for all logical and mathematical concepts, and allows for a clear distinction between the formal notions and the possibility of intuiting them. Jean Cavaillès, the eminent French philosopher of mathematics, also argues that phenomenology and definitude are closely linked. He claims that Gödel's incompleteness results not only preclude acceptance of definitude as characterizing mathematical systems that include arithmetic, but also have consequences for Husserl's phenomenology (Cavaillès 2021, 70–2). According to Cavaillès, these results show that any a priori constraint on mathematics should be rejected. He further claims that phenomenology's approach to logic and mathematics assumes some sort of a priori constraint, in presupposing that it is possible to fully capture these domains.[13] Cavaillès does not explain this claim. One way to interpret it, however, is in terms of the connection Hartimo draws between definitude and categoricity. On this interpretation, for a mathematical notion to be fully captured by an intention, it must be captured by a theory that is categorical, and therefore it must be captured by an axiomatic system that is definite.

3.2 Intuition and Mathematics

According to Husserl, for mathematical concepts to be fully understood, they must be intuited. Understanding that remains at the symbolic level does not suffice. Such intuitions are required, for example, for re-identifying mathematical concepts. For instance, understanding how the symbol "+" is used in a given mathematical formula is not enough: there has to be a possibility of identifying the operation of addition as having a certain characterization, and recognizing it as the same operation that was used previously. This requirement can also be

---

[12] Centrone (2010) and da Silva (2016) argue that the sole role of definitude is to provide something like what is currently called syntactic completeness (although, as Centrone stresses, at the turn of the twentieth century there was no clear distinction between syntax and semantics). I concur with Hartimo's stance, since in my view, Husserl did not take the role of the notion of definitude to be limited to syntactic completeness.

[13] Cavaillès's claim that Gödel's incompleteness results undermine the tenability of phenomenology is not accepted by all. Bachelard (1968) argues that the results indeed undermine Husserl's definition of definitude, but not phenomenology as a whole; Lohmar (2000) argues that because Husserl's definition of definitude isn't a formal definition, it isn't vulnerable to the incompleteness results.

viewed from a non-Husserlian perspective. The need for such an identification is manifest in Kripke's famous interpretation of Wittgenstein's rule-following problem: how can we be sure that the "+" we're using in an arithmetical equation now has the same meaning as the "+" we used in another equation a moment ago? On the Husserlian conception, the assurance in question is provided by an intuition of the meaning of "+". The need for a level of apprehension that secures the identity and objectivity of mathematical concepts exists even in cases where we have a full formal characterization of a concept. Intuitive apprehension helps us understand the meanings of concepts and provides the ultimate justification for what we take them to signify. Another context that sheds light on the role of intuition in our comprehension of mathematical concepts is that of the need to justify rules of inference in logic. The formal modus ponens rule, for instance, does not in itself provide the ultimate justification for the inference from "if p then q" and "p" to "q". Something else must assure us that this rule is correct. Yet another context in which intuition is necessary is that of the justification of a formal system's axioms. An intuitive grasp of the truth of the axioms provides this justification. According to Husserl, "*Immediate 'seeing,'* not merely sensuous, experiential seeing, but *seeing in the universal sense as an originally presentive conscious-ness of any kind whatever*, is the ultimate legitimizing source of all rational assertions" (*Ideas*, §19; italics in the original).[14]

Before going into the various interpretations of this Husserlian notion of intuition in the context of mathematics, it will be useful to compare it with two central positions on the role of intuition in mathematical knowledge, those of Descartes and Kant. According to Descartes, intuition and deduction are the two sources of knowledge in mathematics, and "intuition is the indubitable concep-tion of a clear and attentive mind ... it is more certain than deduction" (Descartes 1985, 14). Intuition is self-evident and certain. It does not require additional justification. Descartes maintains that first principles are known only through intuition.

Kant argues that his thesis that all judgments of geometry and arithmetic are synthetic a priori entails the need for intuition in mathematics (Kant 1998, 143–5). Being synthetic a priori, the truth of judgments of mathematics cannot be solely conceptual, but must be grasped by intuition, specifically, through the pure intuitions of space and time. According to Kant, intuition generates an immediate relation to objects (155), whereas concepts can have only an indirect

---

[14] The notion of intuition as the fulfillment of intention is invoked from *Logical Investigations* onward. Husserl's notion of an authentic presentation of a natural number in *Philosophy of Arithmetic* can be interpreted as an early articulation of the notion of intuition, since it serves the same purpose within his account of number as the subsequent notion of intuition.

relation to objects. Kant takes the distinction between intuitions and concepts to mirror a distinction between two cognitional faculties: sensibility and understanding.

Husserl's notion of intuition is closer to Descartes's position than to Kant's, although it differs from both.[15] For Husserl, intuition is unlike symbolic understanding, which remains at the level of mere words. Intuition is an immediate mode of knowing. Like Descartes, Husserl takes intuition to provide self-evident knowledge. Unlike Kant, Husserl does not take intuition to be grounded in a special faculty. Husserl criticizes Kant's conception as psychologistic.[16] A related difference between Husserl's view and Kant's is that Husserl affirms, and Kant denies, that concepts and categories can both be intuited. According to Kant, concepts can be thought, but not intuited. Furthermore, Husserl claims that Kant does not provide a satisfactory justification of the logical concepts, which is a precondition, he argues, for accepting the forms of judgments. The realm of logic is not adequately captured by the claim that it is subject to the law of contradiction (*LI* VI, §66).

In *Logical Investigations*, the intuition that is relevant to mathematics is categorial intuition, discussed in Section 1. There are two main motivations for adopting the notion of categorial intuition in the realms of logic and mathematics. First, categorial intuition is important in establishing the relation between norms and truth. This nexus is discussed in the context of Husserl's critique of psychologism. If logic is understood only as a technique for arriving at results on the basis of given rules, that is, for deducing conclusions from premises by applying rules of inference, psychologism with regard to these rules has not been eliminated. Consider, for instance, the law of noncontradiction. If we view it as a technique for reaching conclusions, our rationale for adopting it in the first place may well be psychological, namely, the feeling that we are compelled to accept it. Husserl contends that the only justification for adopting the law of noncontradiction, or any other law or rule of logic, is that it is true.

The normativity of the laws of logic and mathematics depends on their truth, and hence, on there being intuitions that fulfill intentions vis-à-vis these laws; such intuitions can only be categorial intuitions (*LI* VI, §62). Furthermore, as we saw in Section 1, the need for something like categorial intuition arises from Husserl's conception of intuition as the fulfillment and confirmation of intention. The fulfillment of the meaning of the claim that there are five things on the table includes the five-ness of the aggregated things (*LI* VI, §40). After *Logical*

---

[15] See Hintikka (2003) for a comparison of Husserl's notion of intuition with the pre-Kantian and Kantian conceptions of intuition.

[16] See Husserl (1956, 402).

Investigations, Husserl abandons the notion of categorial intuition, but he does adduce the closely related notion of seeing essences.[17]

The core of Husserl's conception of intuition is that intuition is the fulfillment of intention.[18] With regard to mathematics, intuition can be plausibly understood in two main ways:

1. As a kind of perceiving: the intuiting of mathematical notions is something like perceiving them. This perception can be interpreted either as akin to sensory intuition or as a "seeing" of essences that need not be akin to sensory intuition.
2. As proof: a proof of a mathematical theorem amounts to an intuition of that theorem.

3.2.1 Intuition Understood as Perception

One of the two plausible interpretations of Husserlian intuition in the mathematical realm regards it as a kind of perception. In *Ideas*, Husserl says that "empirical intuition or, specifically, experience, is consciousness of an individual object. . . . In quite the same manner intuition of an essence is consciousness of something, an 'object'" (§3). Similarly, in *Formal and Transcendental Logic* (§58; italics in the original), he says that "the *objectivity* of something ideal can be directly 'seen' (and, if we wished to give the word a suitably amplified sense, directly experienced) with the same originality as the identity of an object of experience in the usual sense." Already in *Philosophy of Arithmetic*, perception is related to Husserl's notion of an authentic representation: a representation is authentic when we see a thing itself in person, as it were (*PA*, 205). From *Logical Investigations* onward, perception is conceived as fulfillment of intention. This fulfillment is an agreement between what is meant (intended) and what is perceived, and this agreement is truth (*LI* VI, §38). Perception involves the recognition of something as something (e.g., a house as a house) (*LI* VI, §§6, 37). Although Husserl claims that there is perception of universals (*LI* VI, §52), he also says, apropos the founded object, that is, the object at the categorial level, that "the thought of a straightforward percept of the founded object . . . is

---

[17] Husserl does not give any explanation for the change in his view, but it probably has to do with the broader reasons for the shift to transcendental phenomenology: categorial intuition is too close to Platonism vis-à-vis categories, and is not conducive to a clear distinction between psychology and phenomenology. Lohmar (1990) argues that Husserl abandoned the notion of categorial intuition due to its psychological residue. Hopkins (2011, chapter 32) contends that the problem posed by categorial intuition is specifically related to mathematics: categorial intuition cannot adequately ground the realm of formal mathematics.

[18] On the import of construing mathematical intuition as fulfillment of intention, see Tieszen (1989).

a piece of nonsense" (*LI* VI, §61). Hence, while Husserl can be taken to construe the intuition of mathematical notions as in some way analogous to perception, exactly how this analogy works is unclear.

In twentieth-century philosophy of mathematics, the best-known theory on which intuition of mathematical concepts is akin to perception is that proposed by Gödel:

> But, despite their remoteness from sense experience, we do have something like a perception also of the objects of set theory, as is seen from the fact that the axioms force themselves upon us as being true. I don't see any reason why we should have less confidence in this kind of perception, i.e., mathematical intuition, than in sense perception. (Gödel 1964, 268)

Gödel's endorsement of perception in mathematics is related to his Platonism.[19] Platonism is the view that mathematical objects have an independent existence, much as rocks and chairs do, and we can therefore perceive them much as we perceive rocks and chairs. Does Husserl's claim that mathematical notions are intuited have any affinity with Platonism? Answering this question is no easy task. Husserl seeks to distinguish between the realm of meanings and concepts (including mathematical ones) and the natural realm. But Husserl does not base the intuition of mathematical notions on Platonistic assumptions about mathematical notions, and furthermore, as I will soon argue, he contends that the categorial is founded on the empirical (*LI* VI, §60).

Yet Husserl is also opposed to the naturalization of meanings. Therefore, even if mathematical concepts are independent, the sense in which they are deemed independent differs from that in which empirical objects are. Consequently, Husserl would not accept Maddy's naturalistic approach to the perception of mathematical objects (Maddy 1980). Maddy's position is premised on a causal theory of perception that Husserl would have been unlikely to accept. Husserl takes causality to be applicable only in the natural realm, not in the context of relations between intuitions and intentions.

Another problem Husserl considers vis-à-vis the claim that mathematical notions are perceived is that sensory perception does not always provide the exact knowledge that is required for mathematics. Husserl adduces the example of geometrical objects. Consider triangles. They cannot be given in empirical intuition, since empirical triangles do not have the exactitude of geometrical triangles. As Husserl puts it, a geometrical figure is an "ideal limit incapable in principle of intuitive exhibition in the concrete" (*LI* VI, §41). According to

---

[19] Gödel's understanding of perception in mathematics can also be linked to his reading of Husserl. For such an interpretation of Gödel's position, see Hauser (2006). Gödel's interest in Husserl's phenomenology will be discussed in Section 4.

Husserl, "no geometrical proposition holds for the drawn figure as a physical object" (*LI* II, §20).[20]

Yet Husserl claims that "everything categorial ultimately rests upon sensuous intuition" (*LI* VI, §60) and that attribution of the categorial to the empirical – as, for example, in perceiving that a branch is part of a tree, that is, in ascribing the (categorial) part–whole relation to the branch and tree – is founded on the perception of things (*LI* VI, §40).[21] Given that it is problematic to base geometry on sensory perception, are there other mathematical domains that can be grounded in sensory intuition? For example, can the natural numbers be so grounded? It seems that Husserl indeed maintains that intuiting the number of a group of things (an aggregate) is possible. Roughly, his account is as follows. Five-ness is part of our perception of the five things, but it is not perceived in the same manner as sensory perceptions of, say, trees. If the claim that there are five things on the table is fulfilled, we perceive the five-ness in some sense, and this fulfillment is not necessarily based on reflection on the activity of counting.

How does this mathematical intuition work? In *Philosophy of Arithmetic*, a central aspect of the attribution of number is the grouping of things together by conjunction. Is there a difference between Husserl's position in *Philosophy of Arithmetic* and his position in *Logical Investigations*, and if there is, what is the role of categorial intuition in this difference? The following quote from *Logical Investigations* highlights the complexity of this question.

> An aggregate, e.g., is given, and can only be given, in an actual act of assembly, in an act, that is, expressed in the conjunctive form of connection *A and B and C*. ... But the concept of *Aggregate* does not arise through reflection on this act: instead of paying heed to the act which presents an aggregate, we have rather to pay heed to what it presents, to the *aggregate* it renders apparent *in concreto*, and then to lift the universal form of our aggregate to conceptually universal consciousness. (*LI* VI, §44; italics in the original)

---

[20] Tragesser (1989) suggests interpreting the phenomenological notion of intuition in the geometric context as an intuition that something is the case (de dicto), not as an intuition of an object (de re). In other words, in the geometric context, we have a de dicto intuition that *p* (e.g., the proposition that A is an equilateral triangle) is the case, rather than a direct intuition of A's being an equilateral triangle. From the Husserlian point of view, however, the distinction between these two kinds of evidence (de dicto and de re intuitions) is not clear-cut. Pradelle (2020, 38) adduces Husserl's statement: "Predicative includes pre-predicative evidence" (*CM*, §4), arguing that it shows that Husserl does not distinguish between de dicto and de re in general, and in particular, does not do so vis-à-vis intuition.

[21] On Husserl's notion of founding in *Logical Investigations*, see Rota (1989) and Nenon (1997). Nenon distinguishes between two notions of founding in *Logical Investigations*: ontological founding, which is based on the part–whole relations developed in Investigation III, and epistemic founding, developed in Investigation VI. In the present section, the epistemic notion of founding is most germane, but Husserl links the ontological and epistemic senses of founding.

On the one hand, Husserl stresses the centrality of the act of assembling. Since assembling is a psychological activity, this might seem to imply that in *Logical Investigations* Husserl does not abandon the psychological element of the account of natural numbers put forward in *Philosophy of Arithmetic*.[22] But, on the other hand, Husserl explicitly critiques positions that "tried to explain the conjunctive association of names or statements through a mere conjunctive coexistence of nominal or propositional acts" (*LI* VI, §51). Moreover, there seem to be some important differences between the accounts given in *Philosophy of Arithmetic* and *Logical Investigations*. First, in *Logical Investigations*, reflection is barely mentioned in Husserl's remarks on conjunction and the natural numbers. This contrasts with Husserl's account in *Philosophy of Arithmetic*, discussed in the previous section; that account maintained that collective combination is grasped by reflecting on the act in which a multiplicity is unified, and that the concept of "one" is arrived at by a process of abstraction and reflection. In *Logical Investigations*, intuition replaces reflection as the source of the givenness of concepts such as conjunction.[23] Second, in *Logical Investigations*, but not in *Philosophy of Arithmetic*, natural numbers are characterized as ideal species.

In *Logical Investigations*, Husserl does not provide a detailed exposition of his position on the natural numbers, or explain where it diverges from the account in *Philosophy of Arithmetic*, but such an exposition can, I contend, be formulated on the basis of work by Fine.[24] A possible direction for fleshing out Husserl's account was already alluded to by Adolf Reinach, one of the early phenomenologists, whose position was inspired by Husserl's *Logical Investigations* (Reinach 1989). Reinach takes numbers to be objective determinations of things. Husserl's view that intuitions fulfill intentions which ascribe number to collections is in harmony with Reinach's position. But Reinach doesn't explain how these objective determinations are to be understood.

Fine interprets natural numbers as structures of arbitrary objects. For example, the number 2 is a structure of two arbitrary objects that are distinct but connected (Fine 1998, 613). Fine's account is developed as an interpretation

---

[22] Hopkins argues that in *Logical Investigations*, Husserl does not provide a nonpsychological account of conjunction; see Hopkins (2011, chapter 32).

[23] To avoid having to ground the notion of number in psychological capacities, Husserl suggests that the intuitive aspect of attributing number to an aggregate of things can be based on a figural structure of the aggregate, for example, a heap or a row (*LI* VI, §51). Although Husserl had already discussed the notion of figural structure in *Philosophy of Arithmetic* (*PA*, 216), it did not, in my opinion, play the same grounding role there that it plays in *Logical Investigations*. Husserl uses Ehrenfels's notion of "Gestalt" to characterize this figural content.

[24] For a detailed explanation of what such an account would look like, see Roubach (2021).

of Cantor's conception of natural number: "Since every single element *m*, if we abstract from its nature, becomes a 'unit,' the cardinal number $\bar{\bar{M}}$ is a definite aggregate composed of units" (Cantor 1915, 86). This definition is very similar to Husserl's *Philosophy of Arithmetic* conception of number as a collective combination of "one"s.[25] But Fine's position fits even better with Husserl's position in *Logical Investigations*, for three key reasons. First, Fine takes natural numbers to be objective structures of things. Second, these structures are not independent of the domain of things to which they are applied, but are grounded in the things being counted. This is much like Husserl's approach, which takes the categorial to be founded on concrete objects and their apprehension. Third, Fine's theory of natural numbers is linked to phenomenology by the fact that one of Fine's motivations is to provide an account of the way mathematicians speak about arbitrary objects ("let *n* be an arbitrary natural number"), a motivation shared with phenomenology. Interpreting Husserl's position on the basis of Fine's theory, the intuition of a natural number would be an apprehension of a structure that is grounded in the intuition of specific aggregates (e.g., the intuition of the number 2 as a structure is grounded in the intuition of two specific objects, say, a chair and a table).

This reading helps us to elaborate on Husserl's remarks about intuiting essences and universals (*LI* VI, §52). In *Logical Investigations*, the intuition of essences ensues from ideational abstraction. The element of abstraction in the intuition of essences lends itself to interpreting Husserl's approach to number in *Logical Investigations* along the lines of Fine's Cantorian theory, as the movement from an aggregate of particular objects to a structure of arbitrary objects.

Later (e.g., in *Experience and Judgment*, §87), Husserl speaks of "seeing" essences. This seeing is achieved by a process of variation. By varying the perspective from which we view – in our imagination – a specific spatial object, a table, for instance, we discover its essence as a spatial object.[26] Spatial objects are always given only partially. Can such a perspectival approach be applied to intuition in mathematics? It seems plausible to construe seeing part of a mathematical notion as akin to seeing part of an empirical object.[27] On this construal, mathematical notions, like spatial objects, can only be grasped partially. In light of the incompleteness results, this "always only partial"

---

[25] Moreover, Fine's account defends Cantor's conception against Frege's critique. Frege argues that units that are counted cannot ensue from a process of abstraction. As noted in the previous section, Frege's critique is also directed at Husserl's approach to natural number in *Philosophy of Arithmetic* (Frege 1972, 324). Fine does not explicitly link his approach to Husserl's position.

[26] The idea of free variation is mentioned in *Logical Investigations* (*LI* III, §5), but only later (e.g., in *EJ*, §87a) becomes the primary means of seeing essences.

[27] A perspectival conception of mathematical intuition is developed in Tieszen (1989, chapter 6).

approach would also be suitable as an interpretation of Gödel's notion of intuition, which takes the intuition of mathematical notions to be similar to the perception of physical objects.

According to Husserl, essences are characterized as that "without which the object cannot be intuitively imagined as such" (*EJ*, §87). The realm of essences is the realm of pure possibilities (*EJ*, §90). Husserl connects his conception of essences to Plato's, but does not adopt Platonistic metaphysical assumptions about them. Essences do not exist in themselves, separately. Only physical objects exist separately.

Given that the intuition of essences does not assume that they exist independently, there are two principal interpretations of Husserl's view of essences:

1. Essences are dependent aspects of sensory objects. They exist, but not separately. This interpretation, which can be called the Aristotelian interpretation, is close to the Finean interpretation of Husserl's position just proposed. It faces two problems. First, it significantly limits the mathematical domains that can be given in intuition. Only the realms of mathematics that are applicable to sensory reality can be given in intuition. Second, Husserl maintains that meanings are not grounded in the sensory (*LI* VI, §57). Is this position compatible with construing the intuition of essences as the intuiting of aspects of sensory objects?[28]

2. No ontological commitment vis-à-vis essences need be made.[29] This can be called the Humean interpretation. It takes truths about essences to be analogous to Humean relations of ideas. The problem with this interpretation is that it leaves the standing of categorial essences unclear. Moreover, if essences are meanings that do not exist at all, this seems to blur the distinction between intuiting such meanings and understanding them conceptually.[30]

3.2.2 Intuition Understood as Proof

We have examined the first of two plausible interpretations of Husserlian intuition in the mathematical realm, namely, the interpretation that takes intuition to be a kind of perception of mathematical essences. Although, as we saw, it is possible in some cases to relate intuition in mathematics to perception, it is

---

[28] One way to meet this challenge is to claim that though meanings are not grounded in the empirical realm, the intuition of meanings is.

[29] For an example of this sort of interpretation, see Thomasson (2017).

[30] Interpreting Husserl's position as not making a sharp distinction between intuiting a meaning and grasping it conceptually can be seen as reflecting his claim that categorial intuition is the real performance of categorial acts, that is, it is the actualization of categories by thinking with them, rather than a separate way of grasping them (*LI* VI, §62). This interpretation of Husserl's position, which has been suggested by Pradelle (2020, chapter 2), will be discussed below.

difficult to deem perception applicable to all mathematical notions, especially very abstract notions such as large cardinals. We also saw that Husserl rejected the Platonistic interpretation of intuited mathematical essences, but neither the Aristotelian nor the Humean reading of what essences are can provide a satisfactory account of mathematical intuition. We thus now turn to the second option for interpreting Husserlian intuition in the mathematical realm: understanding intuition in terms of *proofs*. This interpretation is not based on the perception model, but rather takes proofs to constitute evidence that a thought or theorem is true. On this approach, proofs are not just signs organized according to a set of rules, but have epistemic import.

Does every proof provide evidence in the required sense? It is true that every proof provides a justification for accepting a theorem, but does this justification always correspond to intuition as Husserl conceives it? It does not, for two important reasons. First, an intuition, as Husserl understands it, seems to be more than a mere justification for accepting a theorem. It is a sort of insight, a way of "seeing" that a theorem is true, but not every proof provides such insight.[31] Second, Husserl argues that intuition provides the ultimate ground for mathematics, but it is not the case that the entire body of mathematics is grasped by intuition. According to Husserl, "*the realm of meaning is, however, much wider than that of intuition*" (*LI* VI, §63; italics in the original). If so, all of mathematics does not meet the condition of being grounded in intuition. But were every proof equivalent to an intuition, all accepted mathematical theorems would be intuitively given.

The idea that mathematical intuitions must amount to more than mere proofs appears to be related to the practice of mathematics. Mathematicians are not always content to have only a single proof of a theorem. From the perspective of mathematical praxis, it would seem that not every proof provides satisfactory evidence for a theorem. Inspired by Husserl, Gian-Carlo Rota has suggested a way to link proofs to intuition without requiring all proofs to be grounded in intuition. Rota ponders why mathematicians put so much effort into proving a theorem that has already been proven. His answer is that they are trying "to debunk the fakery that lies concealed underneath every logically correct proof" (Rota 1997a, 195). According to Rota, this is the main task that occupies mathematicians (Rota 1991, 135). An approach that, though not rooted in the phenomenological tradition, is similar, distinguishes between mathematical proofs that are explanatory and mathematical proofs that are not explanatory,

---

[31] In the text "On the Logic of Signs" from 1890, Husserl puts forward the requirement that genuine knowledge must provide insight into the meaning of a symbolic representation. I contend that this reflects the idea that not every proof constitutes an intuition about a mathematical theorem; see Husserl (1994, 47).

a distinction that goes back to Proclus, if not further. In analytic philosophy of mathematics, systematic articulation of a notion of explanation in mathematics began with the work of Mark Steiner.[32] According to Steiner, an explanatory proof of a theorem is a proof that hinges on the characterizing properties of the entities referred to in the theorem, which he also calls their "essence" (Steiner 1978, 143). Steiner's approach can therefore be seen as a way to interpret Husserl's notion of intuition in mathematics in terms of explanatory proofs.

Another direction for interpreting the notion of intuition in terms of proofs is to argue that it is not a specific kind of proof, or any single proof, that provides intuition into mathematical theorems and notions. Rather, intuition is the gradual enhancement of our understanding of mathematical notions, enhancement attained by having more and more proofs of the theorems and notions in question.[33] Understanding of a mathematical notion is reached by relating it to other notions through proofs. This interpretation meshes well with Husserl's assertions that "it is *essential* to these [categorial, MR] acts, in which all that is intellectual is constituted, that they should be achieved in stages" (*LI* VI, §57; italics in the original) and that there is "graded transition from intention to fulfilment," the final fulfillment being an ideal, namely, the notion's being made self-evident (*LI* VI, Introduction).[34]

3.2.3 Can the Two Understandings of Intuition Be Reconciled?

Is it possible to reconcile the two understandings of the Husserlian notion of intuition, namely, intuition understood in terms of proof, and intuition understood as perception? One option is to interpret proofs as perceptions. This accords with the characterization of perception as dichotomous: either we see the table or we don't. But there is another way of reconciling proofs and perception. The perception of spatial objects always takes place from a certain perspective. This feature can also be adduced vis-à-vis proofs.[35] Any proof captures a mathematical theorem or notion only partially. On this interpretation, the goal of accumulating proofs of the same theorem is to arrive at a fuller view of the theorem.[36]

---

[32] For an overview of the different conceptions of explanation in mathematics, see Mancosu (2018).

[33] See Pradelle (2020) for a development of this interpretation.

[34] On this interpretation of Husserl's notion of categorial intuition, the history of mathematics can be seen as generating successively clearer intuitions of the mathematical concepts. This interpretation thus links Husserl's position in *Logical Investigations* to his later position, expressed, for example, in "Origin of Geometry," in which the history of mathematics plays a central role. Husserl's later position is discussed in Section 3.3.2.

[35] That proofs are related to perceptions in this manner is suggested in Pradelle (2020, 133).

[36] This interpretation of Husserl's notion of intuition meshes well with his position on the Frege–Hilbert controversy. Husserl read the correspondence between Frege and Hilbert regarding geometry, commenting on various points raised (*PA*, 468–73). Frege objected to the idea that it

The analogy between perceptions and proofs can be taken further if we interpret the mathematician's activity of accumulating multiple proofs as analogous to the perceiver's accumulation of kinesthetic experiences, which are, Husserl contends, essential for extraction of the invariant essence of the perceived object (*EJ*, §87).

Yet the attempt to reconcile perception and proofs by invoking their gradual, cumulative character is problematic insofar as it is unclear why we would characterize such processes as intuition. Intuition is ordinarily taken to be an immediate apprehension, and is dichotomous, not a matter of degree: either you have an intuition or you don't. Activity that seeks to achieve a comprehensive understanding of a mathematical notion seems like something quite different.

3.3 Transcendental Phenomenology and Mathematics

The transcendental turn in phenomenology leads to a change of focus: the way things are given is examined from the perspective of consciousness, and ultimately, from the perspective of the transcendental ego. The shift to transcendental phenomenology is also relevant to Husserl's understanding of mathematics. A central element of the turn is Husserl's focus on the distinction and correlation between noesis and noema. This focus is reflected in the phenomenological inquiry into logic and mathematics, each of which has both an objective and a subjective side (*FTL*, §8). The distinction between noesis and noema, and so too, their correlation, renders the mathematician relevant to constituting the meanings of the mathematical notions. After the transcendental turn, Husserl understands his pre-phenomenological account of natural number in *Philosophy of Arithmetic* as a "constitutional" account (*FTL*, §27).

How is constitution to be understood? How is it distinguished from a psychological process? Is it a process in time? Is it identical to construction in time? Is a constituted meaning based on a pre-given ideal? (*FTL*, §73). Can a position based on meanings constituted by mathematicians avoid rendering mathematics subjective? For any given notion, how can we ensure that different mathematicians constitute the same meaning? Is a concept constituted by one

---

is the theory of geometry as a whole that fixes the meaning of the basic notions of geometry (e.g., point, line). For the meaning of a notion to be determinate, Frege argued, the notion must be given an independent definition. Husserl supports Hilbert's conception of a formal system, which is close to his own conception. Moreover, Hilbert's conception is also in harmony with a critique of Frege Husserl voices in *Philosophy of Arithmetic*. Contra Frege, Husserl maintains that the basic logical and mathematical concepts cannot be defined (*PA*, 124–5). In light of the interpretation of intuition that takes it to arise from the totality of proofs of a theory, and not as an independent grasping of the meaning of a basic notion, this could be another point of agreement between Husserl and Hilbert (although Husserl does not say so explicitly). Husserl's position on the Frege–Hilbert controversy is also discussed in Section 4.2.

individual, or intersubjectively? Does the idea that the basic concepts of mathematics are constituted, as opposed to being discovered, entail that mathematics is inherently limited, and if so, does that make transcendental phenomenology a revisionist position vis-à-vis mathematics? These questions will be considered in this section and the following one. We will begin by discussing the constitution of two key mathematical notions: *set* and *continuum*, after which we will present various models of constitution.

3.3.1 Constitution of Mathematical Concepts

Set

According to Husserl, the constitution of a set is not just an apprehension of one object and then another one, such that the previous object is retained in the second apprehension, and so on. Rather, it requires an additional act in which unity is attributed to the objects apprehended. In the case of two objects, in addition to apprehending the objects one after another, there is an additional act, that of grasping the two objects together. A set is constituted by this unifying act of consciousness; it is the "noetic unity of consciousness" (*EJ*, §61). Only afterward, in retrospective apprehension, does this act become an object. A set is defined by its members: "every set must be conceived a priori as capable of being reduced to ultimate constituents" (§61). Sets can be colligated to create sets of sets. With respect to their characterization in language, that is, in judgments, sets are related by means of conjunctions ("and") or disjunctions ("or").[37]

The idea that mathematical notions are constituted raises the question of the relation between the phenomenological notion of constitution and Husserl's earlier notion of the unity of a multiplicity, discussed in *Philosophy of Arithmetic*. At first glance, there do not seem to be any significant differences between the two approaches. Might "constitution" simply be another name for the psychological account of the mathematical notions? This question poses a serious challenge to phenomenology. One possible answer invokes the correlation between noesis and noema: every noesis related to constitution of the notion of set is correlated with an objective noema, hence constitution has an objective as well as a subjective side.

According to Husserl, an advantage of the idea of constituting mathematical concepts is that it enables us to preempt the logical paradoxes, and specifically, those arising from the notion of set, thereby helping to overcome the crisis of foundations in mathematics (*Britannica*, §12). Through constitution, the

---

[37] On Husserl's account of sets, see da Silva (2013).

meanings of mathematical notions, such as that of set, are rendered transparent, and do not rest on any hidden assumptions.[38]

Continuum

The nature of the continuum has engaged mathematics and philosophy since antiquity. The continuum raises two key questions: can a number be assigned to every segment of the continuum (the problem of the irrationals), and is the continuum made up of points? In the late nineteenth and early twentieth centuries, mathematicians proposed new answers to these questions. Along with Dedekind and Cantor, Weierstrass, Husserl's teacher, arithmetized the continuum, that is, described it as a multiplicity of points, each of which can be assigned a number. Husserl discusses the Cantor–Dedekind conception of the continuum in a text from the early 1890s. He ponders whether the Cantor–Dedekind definition of the continuum as a dense set that contains all the limits of the fundamental series in it should be considered a description of a previously given notion of the continuum, rather than a definition in its own right (Husserl 1983, 84). Husserl was also familiar with at least some of Brentano's discussions of the mathematical notion of the continuum.[39] Both in *Philosophy of Arithmetic* and in *Logical Investigations*, Husserl seems to uphold the arithmetization of the continuum. The idea that natural numbers are the basis of mathematics in its entirety, and the conception of mathematics in terms of the theory of manifolds – which, as we saw, Husserl endorsed – seem to support the conception of the continuum as a multiplicity of points, each of which has a numerical value.

Yet there is tension between this arithmetized conception of the continuum and various ideas Husserl came to uphold as his phenomenology developed, particularly that of temporal consciousness. Husserl views consciousness of time as more than just a specific realm of consciousness, for it plays a central role in phenomenology in general. According to Husserl, our basic consciousness of time is not consciousness of a punctual present, but consciousness of an interval that encompasses a retention of the near past, a primal impression of the present moment, and a protention toward the future. In *Ideas*, Husserl questions whether the mathematical account of the continuum as a set corresponds to our experience, asking "Is the stream of consciousness a genuine mathematical manifold?" (§73). Moreover, in *The Crisis of European Sciences*, Husserl explicitly objects to the arithmetization of geometry: "This arithmetization of

---

[38] Husserl discusses the paradoxes of set theory in a text from 1912; see Ierna and Lohmar (2016).

[39] Husserl attended a seminar at which Brentano discussed the notion of the continuum, and refers to Brentano's conception of the continuum in his "Thing and Space" lectures (Husserl 1997, 59). On the influence of Brentanian philosophy of mathematics on Husserl, see Ierna (2017).

geometry leads almost automatically, in a certain way, to the emptying of its meaning" (*Crisis*, §8).

Against the background of the mathematical and philosophical focus on the nature of the continuum in the first decades of the twentieth century, the tensions noted by Husserl prompted much discussion of the relation between the continuum as experienced and the mathematical conception of the continuum. The two principal elucidations of the continuum that invoked phenomenological considerations were those put forward by Hermann Weyl and Oscar Becker.

Weyl on the Continuum

Hermann Weyl, a leading early-twentieth-century mathematician, also worked on the philosophical foundations of mathematics. In 1918 he published an essay examining the relation between the intuitive continuum and the mathematical account of the intuitive continuum. Both Weyl's overall framework (e.g., his characterization of a judgment as a claim about a state of affairs) and his understanding of the intuitive continuum, were influenced by Husserl's positions. According to Weyl, time is the most fundamental experience of the continuum (Weyl 1987, 88). This experience cannot be generated by point-like moments (91). In experiencing time, we have, in addition to an experience of a "now," the experience of something slipping away (92). Weyl asks whether it is possible to mathematize a conception of our experience of time that allows for both the determination of temporal relations (earlier and later) and the comparison of time spans. Having spelled out the conditions under which this mathematization is possible, he argues that, from the perspective of the experienced continuum of time, these conditions cannot be met (91–2). Moreover, points are not experienced as entities that exist in themselves. Hence, formal number systems must be distinguished from the experienced continuum, though there is a relation between them. Weyl's conception of the relation between the formal notion of the continuum (i.e., the mathematical continuum) and the experienced continuum can be interpreted as paralleling Husserl's account of the relation between signification and intuition.

During the 1920s, Weyl's position underwent two changes. In 1921, Weyl drew closer to Brouwer's position on the relation between the intuitive (i.e., experienced) and the mathematical continua (Weyl 1998). Weyl now says that the mathematical continuum is a continuum of becoming. Weyl and Brouwer take the experienced continuum to be a constraint on mathematical conceptions of the continuum, determining which conceptions are acceptable. To reduce the gap between the experienced and mathematical continua, Brouwer developed the notion of choice sequences. This notion injects into mathematics an element of becoming and evolving over time. But Weyl's position is not identical to Brouwer's.

Weyl does not accept the notion of an arbitrary choice sequence as a mathematical notion, and hence, given the constraint of the experienced continuum, takes the possibilities for developing a mathematical account of the continuum to be more limited than Brouwer does.[40]

Later, Weyl distanced himself from both Husserl's phenomenology and Brouwer's intuitionism, preferring Hilbert's formalism. The motivation for this shift was recognition of the fact that intuitionism is too restrictive to accommodate certain areas of applied mathematics, particularly in physics. At this point, Weyl maintains that it is mathematics' *applicability* that renders mathematical formulas meaningful.[41]

The influence of Brouwer's thought on Weyl's evolving conception of the continuum raises various questions. Is Brouwer's concept of the continuum in harmony with the phenomenological conception of the continuum, or incompatible with it? And more broadly, what is the overall relation between phenomenology and Brouwer's intuitionism? To what extent are they in harmony? The latter question will be discussed in Section 4.

Oskar Becker on the Continuum

During the 1920s, Oskar Becker, who studied under Husserl and Heidegger, sought to provide a phenomenological response to the debates about the foundations of mathematics. Becker discusses the notion of the continuum in his 1923 essay "Contributions to a Phenomenological Grounding of Geometry and its Applications in Physics" and his 1927 book *Mathematical Existence*. In the essay, he examines the continuum problem from a Husserlian perspective, but also in relation to the work of Weyl and Brouwer. According to Becker, the problem of the continuum is the problem of arriving at a rational conception of pure intuition (Becker 1923, 397). Becker addresses this problem by distinguishing three approaches to characterizing the continuum as a set of points. The weakest, in his view, is the Cantorian approach that defines sets in terms of their members, and the strongest is the approach based on Husserl's notion of definitude, discussed in Sections 2 and 3.1, which Becker (adducing *Ideas*, §72) defines as the possibility of deciding (*Entscheidung*) the provability of all the propositions in an axiomatic system. Becker delineates a middle position, which he attributes to Russell, that defines a set as the extension of a concept, and contends that adopting this middle position makes it possible to answer certain questions that cannot be answered on the Cantorian approach. To illustrate the differences between the

---

[40] Van Atten, van Dalen, and Tieszen (2002) argue that the disagreement between Weyl and Brouwer over choice sequences is related to Weyl's being inclined toward Husserl's view that mathematical objects are nontemporal.

[41] See, for example, Weyl (1949, part I, chapter 2).

Cantorian and Husserlian approaches, Becker gives the example of Fermat's last theorem and the quadruples that satisfy the equation. On the first approach, what is required is that for any quadruple of natural numbers, we can know if it satisfies the equation or not. The approach based on Husserl's notion of definitude requires that we prove whether Fermat's last theorem is true or not, a much stronger requirement. Becker upholds the Husserlian stance. He argues that from the perspective of transcendental phenomenology, Husserl's notion of definitude must be adopted, because it enables us to determine everything that can be true regarding any object, since every proposition about the object is either provable, or its negation is provable (Becker 1923, 412). He further claims that constitution, too, requires adoption of Husserl's notion of definitude, which is compatible with intuitionism (414), but not with the Cantorian conception of a set.[42] Becker argues that this compatibility with intuitionism ensues from the fact that each element of a Husserlian definite multiplicity is constructed individually.

With regard to the mathematical treatment of the continuum, Becker distinguishes three levels: the morphological, the topological, and the geometrical. The morphological level describes our impressions of the form of the continuum (e.g., a line without breaks). The topological level gives formal characterizations of the continuum (e.g., in terms of connectivity) and provides the means for deploying a kind of "net" over the continuum that can then be used to generate algorithms or calculations (Becker 1923, 424). At the geometrical level, Becker argues, the notion of limit can be introduced (424). The notion of limit makes it possible to bridge the gap between rational processes that are sometimes infinite (i.e., the topological level) and the strictly intuitionistic morphological characterization of the continuum (425).[43]

3.3.2 Modes of Constitution

The aim of constitution is to answer the question of how the cognition of idealities as idealities is possible.[44] Husserl's notion of constitution seeks to allow for a position that does not take the cognition of mathematical notions to

---

[42] Becker takes Brouwer's intuitionistic conception of the continuum to be equivalent to the concept Weyl put forward in his 1921 essay (Weyl 1998), although as mentioned above, there are important differences between the two conceptions. Brouwer had, from the mid-1910s, accepted choice sequences as legitimate mathematical entities, whereas Weyl did not accept choice sequences. Becker discusses the notion of choice sequences in his study of mathematical existence (Becker 1927). He maintains that a phenomenological account of choice sequences requires the conception of time proposed by Heidegger in *Being and Time*, rather than the Husserlian notion of definitude.

[43] In his discussion of space in *Being and Time* (Heidegger 1962, 146–7), Heidegger adopts Becker's distinction between these three levels.

[44] Husserl also calls this inquiry "transcendental criticism of cognition" (*FTL*, §66).

be based on the Platonistic assumption that idealities, including mathematical notions, exist independently, yet does not revert to psychologism. Within the phenomenological tradition, the constitution of mathematical notions has been accounted for in various ways. Two of these modes of constitution will now be explored: a "Kantian" option, on which space and time are the basis for the constitution of mathematical concepts, and a "Leibnizian" option that takes monads to be the basis for the intersubjective constitution of mathematical objects.

Geometry and the Experience of Space

Before examining Husserl's thought on geometry and its relation to space, we should distinguish between the two main conceptions of geometry in his writings. In *Logical Investigations*, under the influence of Riemann and Hilbert, Husserl sees the different geometries as formal theories no different from other formal theories such as arithmetic. The different geometries are axiomatic theories, each of which determines a formal manifold. On this conception of geometry, geometrical figures cannot be given as concrete intuitions (*LI* VI, §41), hence geometry is not related to the experience of space. Husserl puts forward a different conception of geometry in *Ideas*, where he takes geometry to be the a priori characterization of the form of space (*Ideas*, §9).[45] So conceived, geometry differs from formal theories such as arithmetic, because it applies to a specific domain, whereas formal theories apply to any object. In Husserl's terminology, geometry is an a priori material science, whereas arithmetic, for example, is a formal science. This approach has clear similarities to Kant's conception of geometry.[46]

Husserl distinguishes between space as it is experienced and objective space. Experienced space is the space of our perceptual experience of objects. Experienced space is constituted by a distinction between a body (the null-point) as the center of vision and objects that are seen. The movement of one's body changes the perspective from which objects are seen. There are other changes as well: some objects disappear and new ones are perceived. Although experienced space has a privileged point of view (one's body as the null-point),

---

[45] Intimations of this approach can, however, already be detected in some Husserlian writings that predate *Logical Investigations*, which was published in 1900. In these texts, Husserl claims that some experience of space is required for notions such as curvature and dimension. See Husserl (1983, 312–47).

[46] There are, however, some important differences between this conception of geometry and Kant's conception. Husserl, unlike Kant, does not distinguish, vis-à-vis space, between that which is given in intuition and that which is generated by the understanding. Husserl rejects the Kantian distinction between the form and matter of space. For a detailed comparison of the Husserlian and Kantian conceptions of space, see Pradelle (2000).

this point of view is just one of many possible points of view. For an object to be perceived from various perspectives requires, in addition to the possibility of the perceiver's moving and seeing it from different angles, the realization that each such perspective is just one of many possible perspectives. Experienced space assumes multiple perspectives as well as rules governing the transition from one perspective to another. Objective space, by contrast, is not perspectival at all, nor is it based on a privileged point of view (such as one's body) (Husserl 1997, 337).

Husserl's thought suggests three main ways in which the experience of space plays a constitutive role vis-à-vis geometry. One suggestion views geometrical shapes as idealizations of shapes given in experience. This approach posits an epistemic transition from, say, experienced triangles to the geometrical triangle. Granted, there is a gap between the imprecise shapes we experience and the geometrical shapes (*Ideas*, §74). But this gap does not entail that there is total separation between geometrical shapes and concrete sensuous intuitions. Indeed, this is why geometry is a material science, as opposed to a formal science like arithmetic. A second option for relating geometry to the experience of space invokes eidetic reduction, the extraction of essences from the specific contexts in which they are given. We arrive at geometrical notions, such as the notion of the circle, through bodily movement and acts of imagining (*EJ*, §89). This conception of the process that reveals the geometry underlying our experiences is similar to the idea that geometries can be characterized in terms of what is invariant under different permutations of geometrical figures, as proposed by Felix Klein's Erlangen Program.[47]

A third suggestion regarding the constitution of geometry from the experience of space is proposed in "Origin of Geometry," Husserl's last work.[48] "Origin of Geometry" discusses the emergence of geometry as a science. It addresses the historical development of geometry, though Husserl is not concerned with geometry's actual history, but with something that could be called its "transcendental" history, that is, an inquiry into the conditions that were necessary for the emergence of geometry as an exact science of space. According to "Origin of Geometry," the constitution of geometry has several layers. The human activity of measuring provides the foundation for geometry (*Crisis*, 376). The second step is constitution of the ideal meanings of geometry by one individual, a sort of first inventor of geometry, though not necessarily

---

[47] For this sort of interpretation of Husserl's "eidetic reduction" suggestion about the role of experienced space in constituting geometry, see Tieszen (2005).

[48] "Origin of Geometry" is related to the discussion of modern mathematical physics in §9 of *Crisis*, and was probably meant to be part of that section.

a historical "first geometer" such as Thales. Third, geometry is transmitted from one person (presumably, a mathematician) to another and from one generation to another. During this process of transmission, the meanings of the geometrical concepts and theorems are preserved. These meanings thus originate as actual thoughts of some specific individual at some specific time, but subsequently are widely disseminated: "the Pythagorean theorem ... exists only once, no matter how often or even in what language it may be expressed" (357). The persistence of these ideal geometrical objects even when nobody is thinking about them is made possible by written language. Through the written words in which geometry is expressed, geometry accumulates and becomes, as Husserl puts it, "sedimentary," that is, part of the grounding for new generations of mathematicians (361). Through their work, geometers passively reactivate the meanings of the geometrical concepts and theorems; they do not need to actively excavate, as it were, the origins of these entities, though it would be possible for them to do so (364).

Mathematics and Time

While the experience of space is related to specific domains of mathematics – geometry and topology – time can be related to the constitution of all mathematics. Husserl takes this constitution to be based on our consciousness of time. According to Husserl, our primary consciousness of time includes not only the present moment, but also retention of the near past and protention of the imminent future, or expectation (Husserl 1991). Building on this basic experience of time, the notion of objective time is constituted in two stages. The first stage involves the possibility of remembering an event and locating it within a series of events in time. Through such recollection, an order of events is created. The second stage is the creation of a unit that enables us to establish distances in time; these steps yield the objective notion of time.

The two central mathematical notions that can be interpreted as constituted by the experience of time are the continuum and the natural numbers. Our basic consciousness of time, which retains the near past in the present, generates the continuum. The movement of the present into the past is continuous. The natural numbers can be interpreted as generated by the irreversibility of the objective order of time.

Linking mathematics to the consciousness of time marks a change in Husserl's conception of the status of irreal or ideal objectivities in general, and mathematical concepts in particular (*FTL*, §61). Mathematical objects are no longer conceived as a-temporal, as they were in *Logical Investigations*, but rather as omni-temporal. They are omni-temporal by virtue of being inherently related to the structure of time: there is no time at which they do not exist (*EJ*, §64c).[49]

Taking the experience of time as the basis for the construction of mathematical concepts is also at the heart of Brouwer's conception of mathematics. According to Brouwer, the intuition of "two-oneness," which is the intuition of the "falling apart" and unification of the moments of time, is "the basal intuition of mathematics" (Brouwer 1975, 127). Brouwer's position raises the question, which we asked in Section 3.3.1 and which will be explored in Section 4.4, of the extent to which Husserl's phenomenology accords with Brouwer's intuitionism. Here we will only say that if one takes the position that Husserl's notion of the experience of time is the sole basis for the constitution of all mathematical concepts, then it can be argued, as Weyl does, that Husserl's phenomenology is largely in harmony with Brouwer's intuitionism, though there are important points of disagreement (e.g., the unconditional validity of the law of excluded middle is flatly rejected by Brouwer, but not by Husserl). Although Husserl does not restrict the sorts of mathematical concepts that can be constituted, given that consciousness of time is the basis for constituting the notion of the continuum, conceiving the continuum as primarily a set of points seems to be ruled out.

Taking time as the basis for the constitution of mathematical concepts raises the question of the historical development of mathematical concepts. This issue comes to the fore in Husserl's later thought. In *The Crisis of European Sciences*, Husserl argues that the change from ancient to modern mathematics was noncontinuous; in "Origin of Geometry" he explores geometry's emergence as a science. How should these accounts of mathematics' historical development be understood? Husserl rejects two interpretive approaches. On the one hand, he rejects the historicistic approach that takes the meaning of a mathematical notion to be relative to a historical period. Husserl maintains that his critique of psychologism also applies to historicism. On the other hand, he likewise rejects the claim that mathematics is completely independent of history. Husserl does not present a fully developed alternative reading of the mathematics–history nexus. A clue to his view can, however, be found in his use of the notion of "historical a priori" in "Origin of Geometry" (*Crisis*, 373). Two interpretations of this notion suggest themselves: it could refer to the a-historical a priori conditions that govern history in general, or, to the historicity of all a priori conditions. I would argue that Husserl has in mind a hybrid

---

[49] On Husserl's evolving views on the relations between mathematical objects and time, see Lohmar (1993).

interpretation: the historical a priori refers to a priori conditions on events, but these conditions change from one historical period to another, and are thus temporal.

Are there specific reasons why we should adopt a historical approach to the constitution of mathematics? Jacob Klein argued that from a phenomenological perspective, the constitution of formal modern mathematics cannot be a-historical (Klein 1940). The transition from ancient mathematics to modern symbolic mathematics is a noncontinuous jump, Klein asserts, because modern symbolic mathematics is formal, whereas ancient mathematics is not. The transition from the ancient notion of natural numbers as pluralities of units to the modern conception of natural numbers as symbols, for instance, is not continuous. Hence, inquiry into the constitution of mathematical notions mandates examination of the history of their development.

Derrida suggested an additional reason for the adoption of a historical approach to the constitution of mathematics in Husserl's later thought (Derrida 1989). It is related to Husserl's claim in "Origin of Geometry" that written language played a crucial role in rendering the geometrical notions objective. According to Derrida, written language reduces the possibility of directly grasping the original givenness of the geometrical notions. Hence, the constitution of the meanings of geometrical notions is a process that takes place over time, that has a history. It is not possible to bypass the history of geometry and go directly to the original meanings of its notions.[50]

Mathematics and Intersubjectivity

At some point Husserl adopts an intersubjective approach to constitution. Intersubjective constitution can be interpreted in various ways, but in the mathematical context, the most suitable interpretation seems to be that based on Leibniz's notion of the monad, which Husserl adopts in, inter alia, *Cartesian Meditations* (*CM* §§33, 55).[51] Why does Husserl consider a community of monads an apt basis for an intersubjective theory of constitution in general, and for the constitution of mathematics in particular? Each monad has a representation of the entire world, including mathematics, but not every monad is able to articulate this representation. Different monads conceptualize mathematics at different levels, but this does not lead to disagreements over the meanings of the mathematical notions. On a monadological model of intersubjectivity, mathematics is not relative to one subject or monad. Nor does this

---

[50] For a comparison of Klein's and Derrida's respective approaches to the historical emergence of the meanings of mathematical notions, see Hopkins (2005).

[51] A phenomenological interpretation of Leibniz's *Monadology* was also suggested by Mahnke (who studied under Husserl and Hilbert); see Mahnke (1917).

model require any limit on what can be known. As in the case of the transcendental ego, there is nothing that, in principle, a monad cannot know. An additional advantage of a monadological interpretation of intersubjectivity is that the coordination between the monads' different perspectives does not presuppose a realistic interpretation of the mathematical notions. On Leibniz's conception, the world is comprised solely of monads, and mathematical concepts have no existence outside them.[52]

3.4 Phenomenology and the Application of Mathematics

The application of mathematics plays an important role in Husserl's thought about mathematics.[53] Husserl's phenomenology conceives the application of mathematics in two different ways. One conception construes mathematics' application to empirical reality as a matter of the fulfillment of intentions. For example, the thought that there are two cups on the table is confirmed by an intuition of the cups that incorporates application of two-ness to the empirical state of affairs. This conception of the application of mathematics is related to Husserl's understanding of mathematics in terms of formal axiomatic systems; these systems can be applied to the empirical realm. The primary difficulty posed by this conception arises apropos applying geometry to the empirical realm, which lacks the exactness of geometry as a mathematical theory.

A second conception of the application of mathematics, which can be called "application of mathematics first," conceives the application of mathematics as a starting point for the extraction of pure mathematical notions. This conception is already implicit in *Philosophy of Arithmetic*, where the notion of the natural numbers is arrived at by reflecting on the specific multiplicities of things to which it applies. This approach is explicit in Husserl's notion of eidetic reduction: the mathematical notions are extracted from permutations of content given in experience.

But there are also tensions between phenomenology and the application of mathematics, tensions arising from discrepancies between the world as we experience it and the world as it is described by mathematical physics. One example is the divergence between our experience of secondary qualities such as colors and sounds, and the quantitative scientific account of these

---

[52] Adopting the notion of the monad as the basis for an account of the intersubjective constitution of mathematics was also suggested by Gödel. According to Wang, Gödel considered Husserl's transcendental phenomenology to provide the best interpretation of the notion of a monad. For a discussion of Gödel's and Husserl's respective conceptions of monads, see Tieszen (2012).

[53] For example, the question of the application of mathematics is a central motivation for his discussion of imaginary numbers in the 1901 "Double Lecture" (*PA*, 411).

qualities (*Ideas*, §40; *Crisis*, §9i). From a phenomenological perspective, the growing estrangement between our experience and the physical world as described by the natural sciences is a serious problem. Many of our experiences in the world are irrelevant to the natural and mathematical sciences. This alienation undermines the possibility of attaining an ultimate justification of the sciences, since such a justification would have to refer to a conscious subject and the world as she experiences it. This is the crisis Husserl discusses in *The Crisis of European Sciences*. One of the questions asked in *Crisis* is why we accept that mathematics is applicable to the world even in cases where there is tension between its application and our experience. Husserl's answer is that in the modern period, the application of mathematics to the physical world is based on a predetermined mathematization of the world that reflects changes mathematics has undergone. The two main such changes, he asserts, are the arithmetization and formalization of mathematics, which are interrelated. These changes detach mathematical theories from any specific content. For example, arithmetical truths are not necessarily about quantities, but rather, are about relations between "somethings in general." Hence, the applicability of mathematics is no longer limited in scope, as it was in ancient times, but unbounded. Everything is mathematizable. In *Being and Time*, Heidegger proposes a similar conception of a predetermined mathematization of the world (Heidegger 1962, §69b). According to Heidegger, modern mathematical physics is based on a projection of the world as mathematical.

Both Husserl and Heidegger maintained that because modern mathematical physics is at odds with our experiences of the world, there should be a limit to the application of mathematics. But whereas Heidegger was willing to relinquish mathematical physics, arguing that since it rests on more basic ways in which the world is disclosed, mathematical physics is not, strictly speaking, true (Heidegger 1962, 415), Husserl was not (*Crisis*, §9g). Although Husserl did not develop an alternative to modern science's universal application of formalized mathematics, a suggestion as to such an approach can be extracted from "Origin of Geometry." Ancient geometry maintained a relation to the life-world, the world as we experience it, because it was grounded in land measurement practices. All geometries that developed subsequently, including formal geometries, are rooted in ancient geometry, hence modern formal geometries are not completely divorced from the experienced world. This is only a partial solution, and it by no means fully bridges the gap between experience and mathematical physics. For instance, it doesn't bridge the gap between our experience of secondary qualities and physics' account of them. It could even be argued that there is no commonality between geometry as the science of our experience of space, that is, what Husserl calls a "material science", and geometry as a formal science, and the two conceptions are irreconcilable.[54]

The application of mathematics poses an additional challenge to the phenomenological approach to mathematics, a challenge raised by Weyl and Becker in the late 1920s. Some of mathematics' applications to physics involve mathematical entities that appear to have virtually no possibility of being given in intuition. For example, Hilbert spaces, with their infinite dimensions, are applicable to the natural world, but apparently cannot be given in intuition. This led Weyl and Becker to abandon Husserl's phenomenology as the philosophical foundation for mathematics.[55] Whether this conclusion is justified is questionable, since Husserl does not claim that all of mathematics must be given in intuition.[56]

4 Phenomenology and Philosophies of Mathematics

This section examines the relation between phenomenology and the major approaches in philosophy of mathematics. We will begin with approaches that were familiar to Husserl, then proceed to newer approaches that have affinities with phenomenology.

4.1 Phenomenology and Logicism

A good starting point for exploring the relation between phenomenology and logicism is Husserl's critique, in *Philosophy of Arithmetic*, of the account of arithmetic Frege sets out in *Foundations of Arithmetic* (Frege 1980b). The critique most relevant to the logicist program concerns the possibility of defining the mathematical notions. Husserl claims that the basic notions of logic and mathematics, such as the notion of "unit," cannot be defined (*PA*, 124–5). Since logicism is generally characterized as the stance that mathematical notions can be defined in terms of logical notions, this critique indicates that at this early point in the evolution of his thinking, Husserl is opposed to logicism.

Husserl's critique of psychologism, and, at least to some extent, his self-distancing from the account of arithmetic he had presented in *Philosophy of*

---

[54] Invoking Jacob Klein's stance on the incommensurability of ancient mathematics and modern formal mathematics, Hopkins argues that Husserl's suggestions vis-à-vis bridging the gap between the two conceptions fail. For a detailed examination and comparison of Husserl's and Klein's respective stances on symbolic mathematics, see Hopkins (2011).

[55] Webb (2017) explains the abandonment of phenomenology by Weyl and Becker along these lines.

[56] I therefore accept the argument, presented in Mancosu and Ryckman (2002), that concepts such as that of Hilbert spaces do not undermine the possibility of grounding mathematics in phenomenology.

Arithmetic in the wake of Frege's claim that it was psychologistic, led to the emergence of phenomenology in *Logical Investigations*. The extent to which Frege's critique played a role in shaping Husserl's account of arithmetic in *Logical Investigations* has been much discussed.[57] Husserl also seems to have retracted the critique of Frege's position he had voiced in *Philosophy of Arithmetic* (*LI* I, §15, footnote). But do these changes signify an affinity between phenomenology and logicism? To answer this question, consider Husserl's notion of intuition. For Husserl, an intuition is the fulfillment of an intention, and thus intrinsically meaningful. Since it is not the case that such intuitions are placeholders, so to speak, for that which is given but cannot be articulated, Husserlian intuitions do not violate Frege's mandate that we must "prevent anything intuitive from penetrating ... unnoticed" (Frege 1980a, 5). Nevertheless, Husserl's principle that cognition cannot be content with "mere words" and that logical concepts "must have their origin in intuition" (*LI*, Introduction to vol. 2 of the German edition, §2) diverges markedly from the logicist approach, which accepts an account of the mathematical notions that is anchored solely in language.

  Another shift away from the stance taken in *Philosophy of Arithmetic* that is pertinent to the phenomenology–logicism relation is the emerging salience, in Husserl's thought during the 1890s, of the notion of *mathesis universalis*, which links logic and (formal) mathematics. The *mathesis universalis* encompasses the entire analytic sphere, that is, all the formal disciplines (including logic, ontology, and much of mathematics) (Husserl 1975, 28–9). Husserl speaks of "the inner unity of formal logic with pure theory of numbers" (Husserl 1975, 38; see also *FTL*, §26c).[58] But how is this unity to be understood? It has two main aspects. First, both mathematical theories, such as the theory of natural numbers, and logical theories, such as the theory of conjunction, are formal theories. As such, they are part of the *mathesis universalis*. Second, the unity between logic and mathematics is related to the distinction between categories of objects and categories of meaning. Some categories of objects are mathematical (e.g., the natural numbers), others are not (e.g., the category of "unity"), whereas the categories of meaning (e.g., the notion of "proposition") are logical, in the traditional sense of logic. The categories of objects are specified by an axiomatic system governed by the categories of meaning. The categories of objects apply to

---

57 See Section 2. Two seminal discussions of this issue are Føllesdal (1994), which argues that Frege's critique was a significant factor in the change, and Mohanty (1977), which argues that the change in Husserl's position occurred before Frege's review of *Philosophy of Arithmetic*.

58 Husserl also claims that Bolzano, whose position on logic (and in particular, on the ideal status of logical notions, e.g., "proposition," influenced him (Husserl 1975, 37), did not see this inner relation between logic and mathematics (Husserl 1975, 38; *FTL*, §26d). Frege, like Husserl, was influenced by Bolzano's views on logic.

the categories of meaning. For instance, every proposition is a unity. It is due to these various connections between the categories of objects and the categories of meaning that Husserl takes formal mathematics and logic to be so closely linked. Nevertheless, Husserl rejects the logicist reduction of mathematics to logic. He and Frege do concur, however, that logic cannot be algebraized, and both he and Frege are critical of Ernst Schröder's algebraization of logic.

Husserl and Frege are also largely in agreement regarding the distinction between arithmetic and geometry. They concur that arithmetical propositions are analytic, but their notions of analyticity differ. Husserl takes analyticity to be closely tied to formality, whereas according to Frege a proposition is analytic if it can be proved from general logical laws and definitions (Frege 1980b, 4). And both maintain that Euclidean geometry is synthetic a priori, though Husserl does not rule out the possibility of a completely formal approach to geometry.[59]

After the transcendental turn, Husserl's approach to the foundations of mathematics increasingly diverges from Frege's. In particular, Husserl's thesis of the constitution of mathematical concepts by a subject distances his thought from logicism. The gap between phenomenology and logicism is even more pronounced in the phenomenologically oriented approaches of thinkers such as Weyl and Becker.

4.2 Phenomenology and Formalism

Formalism seems to be an approach to the foundations of mathematics that is relatively distant from phenomenology. The core tenet of formalism is that mathematical notions have no intrinsic meaning; they are mere symbols, and do not represent concepts or objects outside the theory in which they appear. This position is captured by Hilbert's maxim: "It must be possible to replace in all geometric statements the words point, line, plane, by table, chair, mug" (Reid 1970, 57). Formalism can also be characterized as an approach on which intuition plays no role in mathematics. If we take Husserl's phenomenological approach to require that, although at some level the mathematical notions function as symbols, we must nonetheless have some sort of intuition of at least some of the mathematical notions, then that approach is patently incompatible with strict formalism. Yet of all the approaches to the foundations of mathematics Husserl was acquainted with, he took Hilbert's formalism to have the strongest affinity with his own orientation.[60] He concurred with Hilbert's characterization of mathematics as formal, and also agreed that the process of

---

[59] The separation of arithmetic and geometry can be viewed as characteristic of logicism; see Reck (2013).

[60] For example, in a note in *Ideas*, §72.

formalization is at the heart of mathematics' historical development. In his words, "the tendency to formalization is inherent in arithmetic" (*PA*, 417).

Is it possible to reconcile this apparent tension between Husserl's phenomenology and his explicit support for Hilbert's conception of mathematics as a formal discipline? This question can be addressed in different ways. We can look at the specific elements of formalism that Husserl accepts, and examine whether they are compatible with phenomenology. Alternatively, we can consider the question more generally, and investigate whether an approach such as Husserl's, which takes intuition to play a key role in attaining mathematical knowledge, is compatible with formalism. In Section 4.2, we will have recourse to both strategies for answering this question, keeping in mind that in the 1920s, Hilbert maintained that establishing the foundations of mathematics requires some form of intuition (Hilbert 1964, 142).

In the early 1900s, Husserl's approach is to a large extent aligned with Hilbert's. Husserl endorses Hilbert's stance in the Frege–Hilbert controversy, and takes his own notion of a formal system's definitude to be similar to Hilbert's notion of the completeness of an axiomatic system (*PA*, 436; *Ideas*, §72). There are, however, aspects of Hilbert's position that do not seem to be in line with Husserl's position during this period. For example, Husserl appears to accept the idea that the basic logical notions have content, and does not consider mathematics to be solely a matter of the manipulation of symbols.

To establish the exact relation between Husserl's phenomenology and Hilbert's formalism, two questions must be addressed. First, are there affinities between intuition, as Husserl understands it, and Hilbert's early formalism? Second, is Husserl's notion of the definitude of formal systems inextricably linked to his phenomenology?

Regarding the first question, it is useful to examine Husserl's comments on the Frege–Hilbert correspondence about the nature of mathematical theories, which focused on geometry (*PA*, 468–73). Frege's main claim is that the meanings of mathematical concepts are prior to the axioms in which these concepts are invoked, a priority that is manifested, for example, in the fact that theories must start by defining the concepts they invoke. Hilbert, on the other hand, argues that the concepts of point, line, and plane do not have fixed meanings. Their meanings are determined by the entirety of the axiomatic system that invokes them, and they can have different meanings depending on the domains to which they are applied. It might be argued that Husserl's approach has affinities with aspects of both the Fregean and the Hilbertian views. Like Frege, Husserl contends that we cannot be satisfied with a merely symbolic understanding of the logical and mathematical notions, and insists that one of phenomenology's aims is to achieve clarity about their meanings. And like Hilbert, Husserl argues that it is the axiomatic system as a whole that determines the characterization of a notion it invokes (e.g., the notion "+"). Ultimately, however, I would argue that Husserl is closer to Hilbert than to Frege, because he does not take the meaning of a mathematical notion to be prior to the formal system that invokes it. On the contrary, Husserl regards the formal system as the basis for determining a notion's meaning. Indeed, the givenness in intuition of the mathematical concepts presupposes the formal systems that invoke those concepts.[61]

In seeking to ascertain whether the role of the definitude of formal systems in Husserl's phenomenology attests to a strong bond between phenomenology and Hilbert's conception of mathematics, the salient question is whether Husserl's characterization of formal systems as definite is integral to the phenomenological account of mathematics. In Section 3.1, I argued that there is indeed reason to see the notion of definitude, and specifically, the categoricity of the mathematical notions, as vital for a phenomenological account of mathematics. Husserl takes definitude to be the ideal, the standard that mathematical theories aspire to (*FTL*, §31). Ongoing movement in the direction of formalization is, he maintains, an essential feature of mathematics. Hilbert's conception of mathematics is a realization of this process of increasing formalization. But Gödel's incompleteness theorems having shown that definitude cannot be achieved (in theories that include elementary arithmetic), what is the status of the ideal of definitude? One answer, proposed by Bachelard, among others, is that definitude can continue to be an ideal for mathematics (Bachelard 1968, 53). Although it cannot be actualized, it directs the development of mathematics. Definitude is therefore an ideal in the Kantian sense. Adopting this approach, Husserl's position can be interpreted as taking Hilbertian formalism to be the paradigm for mathematics.

What about the role of intuition in phenomenology? Can it be reconciled with a formalistic interpretation of mathematics, or is it merely a vestige of early mathematics that will be cast off as the formalization of mathematics advances? In Hilbert's thinking around the turn of the twentieth century, the formal axiomatic systems that govern the characterization of mathematical notions do so exhaustively, leaving no role for intuition. But Hilbert's program for the foundations of mathematics from 1917 onward takes a different stance, affirming that mathematics requires some form of intuition. In "On the Infinite" from 1925, for instance, Hilbert states that mathematics cannot be grounded solely in logic, but requires "extralogical concrete objects which are intuited as directly

---

[61] I therefore disagree with Pradelle's argument that Husserl's notion of intuition renders Husserl's position closer to Frege's than to Hilbert's; see Pradelle (2020, 156).

experienced prior to all thinking" (Hilbert 1964, 142). What we intuit are the structures of the concrete symbols of mathematics, not some alleged underlying meanings attached to them. Mahnke, who studied under both Husserl and Hilbert, claims that this stance is phenomenological in spirit. Mahnke concurs with Hilbert that the mathematical *symbols* are intuited, but proceeds to assert:

> However, Hilbert is mistaken when he concludes therefrom that these "signs themselves [are] the objects of number theory," and do not "mean" anything else. On the contrary, the actual sign 1 + 1 has a meaning, just as does the abbreviation 2 according to Hilbert himself: the former sign is the sensible representation of a purely logical relation, the aggregate connection. The true objects of arithmetic *are* simply these *logical* relations; the real value of the numerals, on the other hand, consists in the fact that they are the most suitable sensory bases of the categorial *intuitions* of numbers, and with this, of all that is genuinely arithmetically "evident" [*Evidenzen*]. (Mahnke 1966, 81–2; italics in the original)

According to Mahnke, Husserl and (post-1917) Hilbert concur that intuition is necessary, but disagree about what it is that is intuited. Mahnke disputes Hilbert's claim that it is only the mathematical symbols that are intuited, arguing, in line with Husserl's view, that the meanings of mathematical notions (e.g., the meaning of "+") are also intuited. But as we saw in Section 3, Husserl's understanding of intuition in mathematics can be interpreted in various ways. One such interpretation takes mathematical intuition to be analogous to sensory perception. This interpretation can help bridge the gap between Hilbert's and Husserl's respective notions of intuition, since the sensory intuition posited by Husserl can be construed as intuition of the structures of mathematical symbols.[62]

4.3 Phenomenology and Platonism in Mathematics

Platonism in mathematics is the position that mathematical concepts exist independently as abstract objects.[63] Several arguments can be made for adopting a Platonistic interpretation of Husserl's position on mathematical concepts (in particular, the position taken in *Logical Investigations*). First, assuming, as Husserl does, that mathematical notions are given in intuition, the interpretation that takes mathematical intuition to be analogous to perception arguably implies that the intuited notions exist independently just as perceived objects exist independently. Second, Husserl criticizes empiricistic and nominalistic approaches to general concepts, including mathematical concepts, and argues for the objectivity of these concepts. Not only is it not the case that mathematical

---

[62] Parsons's notion of mathematical intuition (e.g., in Parsons 1980) seems to be a framework that can comfortably accommodate both the Hilbertian and the Husserlian notions of intuition.

[63] See, for example, the definition proposed in Linnebo (2018).

concepts arise from a process of abstraction on the basis of similarity, as the empiricists and nominalists claim, but it is the other way around: recognizing similarity requires an objective common feature (*LI* II, §§3–4). As a ground for this objectivity, independent existence is often ascribed to the concepts in question. Third, phenomenology seeks to remain true to the way things are experienced. Mathematicians often hold Platonistic views about mathematical concepts, and if the goal of phenomenology is to study these views, and not to impose external constraints on how mathematicians conceive their subject matter, then phenomenology should understand mathematics Platonistically.

Some of Husserl's own assertions appear to support a Platonistic interpretation of his position: "Numbers – in the ideal sense that arithmetic presupposes – neither spring forth nor vanish with the act of enumeration, and . . . the endless number-series thus represents an objectively fixed set of general objects, sharply delimited by an ideal law, which no one can either add to or take away from" (*LI* I, §35). The "concepts, propositions and truths" of logic are "an ideally closed set of general objects, to which being thought or being expressed are alike contingent" (§35).

On the other hand, Husserl repeatedly claims that it is not the case that essences exist independently as do natural objects. He objects to the hypostatization of essences (see, e.g., Husserl 1975, 25), and distinguishes between the mode of being of individual natural objects, such as individual people, and the mode of being of individual numbers (e.g., 2) (*LI* I, §31), although he does seem to accept that both these sorts of things (i.e., natural objects and numbers) exist in some way (*LI* II, §§1–2). The question of the independent existence of objects in general, and mathematical objects in particular, gets complicated if we deem evidence to be the sole acceptable ground for asserting this independent existence. From the perspective of evidence, there is no justification for the claim that any entities exist independently. This is why, from a phenomenological point of view, there is no justification for the Kantian distinction between phenomena and things in themselves. The argument from evidence appears to undermine the idea that mathematical objects exist independently.

However, Husserl does say: "There are therefore countless meanings which . . . are merely possible ones, since they are never expressed, and since they can, owing to the limits of man's cognitive powers, never be expressed" (*LI* I, §35). That is, these meanings are independent of human cognition. As we just saw, on some readings of Husserl's thought, it could be argued that Husserl would accept that mathematical notions satisfy Linnebo's three criteria for Platonism in mathematics: existence, abstractness, and independence (Linnebo 2018). Should this Platonistic position indeed be ascribed to Husserl after the transcendental turn? In other words, is constitution compatible with Platonism? I would argue that although transcendental phenomenology can be interpreted as maintaining that abstract objects exist, and in that sense can be construed as realist vis-à-vis mathematical entities, it is not Platonistic in the sense of claiming that mathematical objects exist independently of their apprehension.[64]

4.3.1 Gödel and Phenomenology

Along with the Platonistic conception of mathematics Gödel adopted in the wake of his incompleteness results, his views on phenomenology are often adduced in support of the Platonistic interpretation of Husserl's position. Beginning around 1959, Gödel viewed Husserl's phenomenology as a potentially apt philosophical framework for post-incompleteness mathematics.[65] From the Gödelian perspective, the advantage of the phenomenological framework is that it enables the meanings of mathematical notions to be clarified without recourse to definitions and axioms (Gödel 1961, 383). This enables mathematicians to work with notions that have an objective meaning, yet this meaning is not exhausted by any specific axiomatic system. So understood, the phenomenological approach to meaning seems to support optimism that all problems in mathematics can be solved, or as he puts it, that "every clearly posed yes-or-no question is solvable" (385). Gödel's optimism is grounded in mathematicians' achievements thus far, and their ability to discover new characterizations of the mathematical notions. In terms of Husserl's exposition of phenomenology, this optimism reflects the noema–noesis correlation: all noemata, all meanings, are arrived at by acts of cognition on the part of mathematicians. Noemata are inherently accessible to cognition; there are no limits on mathematical cognition. Moreover, as we saw in Section 3.3.2, Husserl endorsed the idea of looking at mathematical knowledge intersubjectively, that is, in terms of the community of mathematicians, as opposed to individual mathematicians; such a model is suggested, for example, in *Cartesian Meditations*. The phenomenology–monadology nexus was also developed in Mahnke (1917). Gödel similarly took Leibnizian monads to provide a model for the intersubjective approach.[66]

From the Husserlian perspective, how justified is Gödel's appropriation of Husserl? Gödel's reading of Husserl downplays the affinities between Husserl's

---

[64] Rosado Haddock (2010) argues that Husserl's position was Platonistic in *Logical Investigations* and remained so after the transcendental turn; Tieszen (2010) argues that after the transcendental turn, Husserl's position was "constituted realism." I consider Tieszen's account more accurate.

[65] Gödel only writes about Husserl's phenomenology in one place (Gödel 1961), but from his conversations with Hao Wang, and Wang's notes of the conversations, we know that he took a deep interest in it. On Gödel's study of Husserl's thought, see van Atten and Kennedy (2003), Hauser (2006), Tieszen (2011).

[66] See Tieszen (2012) on Gödel's receptiveness to Husserl's monadological model of mathematical cognition.

phenomenology and Hilbert's formalism. Specifically, the links between phenomenology and the notion of a definite manifold must be disregarded. Gödel was familiar with this notion, but presumably did not take it to play a significant role in the phenomenological approach to mathematics. In addition, the Gödelian reading stresses Husserl's affirmation of the role of intuition in mathematical knowledge. As we saw in Section 4.2, in light of Husserl's affirmation of intuition, a strictly formalistic account of mathematics is, from the Husserlian perspective, inadequate.

In assessing Gödel's interpretation of Husserl, another important question to consider is whether this reading is in line with Gödel's own Platonistic outlook, or marks a change in that outlook. Two different answers have been given. One emphasizes Husserl's notion of categorial intuition, and takes it to parallel Gödel's claim that mathematical concepts such as "set" are perceptible just as empirical objects are perceptible (Hauser 2006). Those who give this answer see Husserl's phenomenology as a variant of the Platonistic approach to the foundations of mathematics. The other answer emphasizes the differences between the Platonism of Gödel's earlier thinking and the phenomenological approach he adopted later. This evolution is attested to by the fact that Gödel was drawn, not to the approach Husserl took in *Logical Investigations*, which, as we saw in Section 4.3, can be interpreted Platonistically, but to Husserl's transcendental phenomenology. Tieszen argues that both phases in Gödel's thought are Platonistic, but there was a shift in the sort of Platonism Gödel espoused. He terms Gödel's later position "constituted Platonism," distinguishing it from "simple Platonism" (Tieszen 2010, 2011). Whereas simple Platonism upholds the independent existence of mathematical entities, constituted Platonism emphasizes mathematicians' ability to arrive at answers to all mathematical questions. Transcendental phenomenology, which "brackets out" assumptions about entities' existence, is an apt framework for constituted Platonism. The change in Gödel's position, on Tieszen's interpretation, is a move from Platonism to realism regarding mathematical notions.

I would like to suggest another sense in which Gödel's interpretation of phenomenology can be viewed as faithful to Husserl. The role Gödel assigns to phenomenology in grounding mathematics meshes well with the approach to the meaning of mathematical concepts Husserl takes during his *Crisis of European Sciences* period, especially in "Origin of Geometry." Husserl contends that the constitution of meanings, including the meanings of mathematical concepts, has an essential historical dimension. There is no point at which the meaning of these concepts has been exhaustively established. Were exhaustive exposition of meanings possible, that would open up the possibility of developing axiomatic systems that fully capture the concepts in question, a possibility

Gödel rejects due to his incompleteness results. Yet historical constitution is not contingent: it is constrained by the concepts' original givenness. This is the Platonistic (or more precisely, realistic) dimension of Husserl's outlook in "Origin of Geometry," and Gödel's view is quite compatible with it.

4.4 Phenomenology and Brouwer's Intuitionism

During the late 1910s and the 1920s, intuitionism was considered, within the phenomenological movement, the most promising approach to the foundations of mathematics. Both Weyl and Becker maintained that there was a close connection between intuitionism and phenomenology.[67] During this period, the issue was not so much Husserl's own position, but whether phenomenology in general should side with Brouwer in the debate with Hilbert over the foundations of mathematics. Heidegger's early thought, especially *Being and Time*, which had some affinities with intuitionism, provided additional impetus for exploring the phenomenology–intuitionism nexus.[68] Over the past twenty years, largely due to the work of van Atten (2007, 2010), there has been renewed interest in possible affinities between phenomenology and intuitionism. Van Atten argues that there are indeed strong links between Husserl's views on the foundations of mathematics, especially his genetic phenomenology, and intuitionism. Phenomenology, he asserts, provides good philosophical justification for intuitionism (van Atten 2007, 75).

The importance that the phenomenological account of mathematics ascribes to what mathematicians experience is the primary basis for the claim that phenomenology and intuitionism have a strong affinity. But this claim can be countered by invoking other considerations, discussed above vis-à-vis arguments for and against phenomenology's affinities with formalism and Platonism. To assess the matter, we will examine suggested links between phenomenology and key features of intuitionism.

4.4.1 Phenomenology and the Primordial Intuition of "Two-Oneness"

The salience of intuition in Husserl's phenomenology already suggests an affinity with Brouwer's position, but do the respective concepts of intuition share more than a name? According to Brouwer, the basic intuition underlying mathematics is the experience of time: "the falling apart of moments of life into qualitatively different moments, to be reunited only while remaining separated" (Brouwer 1975, 127). Brouwer calls this intuition "two-oneness." This

---

[67] See Weyl (1928).

[68] On affinities between *Being and Time* and Brouwer's intuitionism, see Roubach (2008, chapter 2). *Being and Time*'s relation to the philosophy of mathematics is discussed below.

characterization of mathematics' core intuition seems patently different from Husserl's notion of intuition as fulfillment of intention, which is not limited to one specific experience. But Husserl's understanding of the experience of time in terms of the present's retention of the near past is actually quite similar to Brouwer's understanding of the experience of time. And this similarity runs deeper: both Husserl and Brouwer link the experience of time and the series of natural numbers. According to Husserl, the experience of time is the basis for the objective concept of time, which takes time to be closely connected to the numbering of moments.

But for the purpose of evaluating whether there is an affinity between Husserl's and Brouwer's respective concepts of intuition, the key question is whether Husserl, like Brouwer, bases all intuitions, especially those related to mathematics, on the experience of time. With regard to the early stages of his phenomenological thought, the answer is definitely no. Categorial intuition, for instance, is constrained neither by time in general nor by any specific experience of time. But the genetic phenomenology Husserl developed later puts forward a conception of the constitution of objects that characterizes all objects, including mathematical objects, in terms of time. Mathematical objects are omni-temporal (*EJ*, §64 c), not a-temporal. This linkage between mathematical objects and temporality is indeed a feature of both Husserl's and Brouwer's approaches, but there is an important difference between their respective positions on the relation between time and mathematics. Brouwer's conception of mathematical objects as temporal is essentially developmental in nature, whereas Husserl's is not. The omni-temporal conception of mathematical entities construes them as linked to the structure of time, but not as developing over time. Becker's 1927 book on mathematical existence (published in the same volume of the *Yearbook for Phenomenology and Phenomenological Research* as *Being and Time*) argues that Brouwer's approach to mathematics has a stronger affinity with Heidegger's approach in *Being and Time* than with Husserl's approach. Heidegger argues that all entities ordinarily taken to be nontemporal, including numbers, are in a sense temporal (Heidegger 1962, 39–40).

4.4.2 Choice Sequences and Phenomenology

In intuitionism, the notion of "choice sequences" is at the heart of the relation between mathematics and time. In the context of intuitionistic set theory, Brouwer develops the notion of choice sequences to provide a mathematical account of the continuum. A choice sequence is a sequence generated by choosing numbers or numerals. For example, we can create the decimal number 0.35211 ... by a sequence of choices (0, 3, 5, 2, 1, 1 ...). At each step, a numeral is chosen. The choice can be based on a rule (e.g., "choose 2 at each step"), but need not be. Choice sequences are thus characterized by temporal development. Brouwer uses choice sequences to define points on the continuum. These points are sequences of nested intervals whose borders are chosen by a cognizing subject. From the perspective of classical mathematics, it is difficult to accept choice sequences as mathematical objects, for two reasons: they are incomplete objects that develop in time, and their development depends on a subject.

These features of choice sequences can also be seen as problematic from the phenomenological perspective. There is, for instance, evident tension between the first feature – that they develop over time – and both Husserl's early characterization of mathematical objects as a-temporal and his later characterization of them as omni-temporal. And the fact that choice sequences are dependent on a choosing subject can be interpreted as dangerously similar to psychologistic approaches on which mathematical or logical concepts depend on an empirical subject.

Can phenomenology nonetheless serve as a philosophical framework within which choice sequences are acceptable? The phenomenological tradition has suggested various answers to this question.[69] Weyl, who sought to demonstrate the compatibility of phenomenology and intuitionism, maintained that choice sequences are not mathematical objects (Weyl 1998, 94). Becker accepted choice sequences as mathematical objects, but argued that the right philosophical framework for them was Heidegger's "historical time," not Husserl's account of our consciousness of time. Heidegger distinguishes between natural time and historical time.[70] Natural time is based on the possibility of iteration; it is time as measured. Historical time, on the other hand, is best grasped by looking at the phenomenon of the passage from one generation – not in the biological, but in the cultural sense, as in "the generation of 1914" or "Gen Z" – to another. According to Becker, the emergence of a new generation is not determined by the previous generations. That is, in historical time, transitions from one period to another are not rule-governed. Applying this distinction to the mathematical realm, a rule-governed sequence such as the natural numbers exemplifies natural time, whereas a non-rule-governed choice sequence exemplifies historical time (Becker 1927, 229). From the perspective of this Element, the relevant question is whether the distinction between natural time and historical time is in harmony with Husserl's phenomenological outlook. I see no reason to think it isn't. Husserl interpreted the experience of time in terms of

---

[69] Van Atten (2007, chapter 5) argues that choice sequences are compatible with Husserl's phenomenology.

[70] On Heidegger's conception of historical time, see Heidegger (1962, part 2, chapter 5).

objective time – Heidegger's natural time – but it can be interpreted in terms of historical time as well.

We will return to the question of the relation between *Being and Time* and mathematics in Section 4.5.1.

4.4.3 Law of Excluded Middle

One of the main tenets of intuitionism is rejection of the unqualified endorsement of the law of excluded middle, the law asserting that any meaningful proposition must be either true or false. In mathematics, this law is assumed, for example, in proofs by contradiction (*reductio ad absurdum*). Husserl seems to accept the law of excluded middle, and it therefore seems that his position is irreconcilable with intuitionism. But this conclusion is not as straightforward as it initially seems. First, there are places where he appears to raise doubts about the excluded middle (*FTL*, §§77, 79).[71] Second, it could be argued that the law of excluded middle is a dogmatic residue that should not necessarily be retained, and there is reason to doubt its applicability, since we do not have adequate evidence for it. More specifically, Husserl equates truth with evidence, hence it could be argued that his notion of truth is manifestly similar to the notion of verification.[72] If the Husserlian notion of truth is understood as verification, then there is no justification for assuming that every proposition is either true or false regardless of our access to it.

In *Being and Time*, Heidegger puts forward an understanding of truth that parallels the notion of truth as verification. Heidegger argues that correspondence theories of truth, including the Husserlian variant, presuppose both truth as an entity's uncovered-ness, and, someone who uncovers that entity – Dasein (Heidegger 1962, 260–1). According to Heidegger, the law of excluded middle is not valid regardless of Dasein's acts of uncovering: "Before Newton his laws were neither true nor false" (269). Heidegger maintains that the Husserlian concept of truth entails rejection of the absolute validity of the law of excluded middle. With respect to the foundations of mathematics, this position leads to the conclusion that Husserl should come down on the side of Brouwer.

---

[71] Husserl's position on the law of excluded middle has been extensively discussed since the 1980s. Tragesser (1984, 2) claims that Husserl seeks to purge logic of the assumption of the law of excluded middle. Mohanty rejects this claim (see, e.g., Mohanty 1991, 102), arguing that Husserl, being committed to decidability in mathematics, accepts the law of excluded middle. Lohmar (2004) argues for a position between those of Tragesser and Mohanty, noting that in *Formal and Transcendental Logic*, Husserl says that although the law of excluded middle is not a purely logical principle, he does not reject it, because mathematicians often rely on it in practice.

[72] See, for example, Parsons (2012, 205).

Yet this conclusion may be too hasty. First, it could be argued that the meaning of truth encompasses the law of excluded middle. Second, it could be argued that much as it can be claimed that, from a phenomenological perspective, there is no justification for accepting the unlimited applicability of the law of excluded middle, so too it can be claimed that there is no phenomenological justification for rejecting it. The second argument, which is weaker than the first, may suffice, that is, the intuitionistic approach may require no more than agnosticism regarding the law of excluded middle.

4.4.4 Is Phenomenology a Revisionist Position?

The discussion about the excluded middle raises the question of whether phenomenology is a revisionist position. Revisionism vis-à-vis mathematics is the position that some of mathematics' accepted norms, such as the unrestricted use of *reductio ad absurdum* proofs, should be modified. Is Husserl a revisionist regarding mathematics? With respect to *Logical Investigations*, the question can be answered in the negative, since Husserl supports a division of labor between philosophy and mathematics, meaning that philosophy does not interfere with the way mathematics is practiced by mathematicians (*LI*, Prolegomena, §71). But vis-à-vis his later thought, the answer is less clear. As noted, Husserl had reservations about the status of the law of excluded middle; he also hoped that phenomenology could help elude paradoxes in mathematics (*Britannica*, 32). These two positions can be taken to lend support to a revisionist interpretation of Husserl's position.

Van Atten distinguishes between weak revisionism, which critiques specific parts of mathematics, and strong revisionism, which accepts theorems that are not accepted by classical mathematics (van Atten 2007, 53).[73] According to van Atten, Husserl's phenomenology can be considered both weak and strong revisionism (54). He argues that the weak revisionism is exemplified by Husserl's claim that phenomenology could help elude the paradoxes of set theory, and the strong revisionism is exemplified by Husserl's claim that the phenomenological conception of mathematical objects as constituted is not constrained by mathematical practice (59). More generally, van Atten takes Husserl's revisionism to reflect the subordination of ontology to phenomenology (57). He therefore rejects the interpretation of Husserl's position that sees it as basing ontology on the *mathesis universalis*, understood as independent of phenomenology. In my opinion, with regard to the approach Husserl takes in *Formal and Transcendental Logic*, the claim that ontology is subordinate to phenomenology is debatable, since it can be argued that formal ontology is prior

---

[73] Van Atten claims that in the weak sense, even *Logical Investigations* is revisionist.

to the phenomenological constitution of the meaning of the notion of being, but it is indeed correct vis-à-vis Husserl's genetic phenomenology (e.g., *Experience and Judgment*), where the priority of formal ontology is not assumed.

Van Atten's revisionist interpretation is contested by many Husserl scholars. Some of these critiques rest on the affinities between Husserl's views on the foundations of mathematics and formalism (Hill 2010); others, on the affinities between Husserl's views and Platonism (Rosado Haddock 2010). There is also textual evidence that Husserl's position is not revisionist. For example, Husserl asserts that all the sciences, including logic and mathematics, are "self-sufficient" (*FTL*, §67). At the same time, it must be conceded that Husserl also maintains that critical examination of the sciences, including logic and mathematics, is necessary, and the sciences should not be accepted as they are without scrutiny of their foundations (§71). One option for reconciling this apparent tension in Husserl's position, suggested by Hartimo (2021), is to interpret the requirement that mathematics be critically scrutinized as intended to evaluate whether or not mathematics as practiced can achieve its presumed goals, and not as intended to bring about revisions in mathematical theory and practice.[74]

4.5 Mathematics, Phenomenology, and Ontology

The conflicting ways in which phenomenology is related to the various approaches to the foundations of mathematics reflect differing responses to the more general question of the relation between phenomenology and ontology. Is phenomenology based on the assumption that entities given in intuition exist independently, or does it consider this assumption unjustified, since we do not have, and as a matter of principle cannot have, evidence for such existence? In other words, is phenomenology closer to the realist or idealist position regarding the existence of objects, or is it ontologically neutral? Of course, this raises the question of what we mean by "existence," of what we mean when we speak of the being of entities. This question is phenomenological, since entities are given as entities and therefore the meaning of "being an entity" is in some way given to us. One answer is that what characterizes entities as entities is that they are given to consciousness. This answer highlights the affinity between phenomenology and the Berkeleyan claim that to be is to be perceived.

---

[74] In addition to the positions discussed in this section (4.4), which focuses on phenomenology's affinities with intuitionism, another approach to the phenomenology–intuitionism nexus has been put forward by Carl Posy. Posy invokes Husserlian phenomenological notions (e.g., the notion of 'horizon') to interpret intuitionism, although he emphasizes that there are important differences between the two philosophical orientations (Posy 2020, 76–83).

But Husserl puts forward a different answer: the most basic characterization of entities is provided by formal ontology, which is part of the *mathesis universalis* and hence related to formal mathematics. This answer raises an additional question that refines the general question, just posed, of the relation between phenomenology and ontology: what is the relation between phenomenology and formal ontology? In the literature, two opposed answers have been suggested. One takes formal ontology to be subordinate to phenomenology, the other takes formal ontology to be an independent discipline that, though indeed connected to phenomenology, is not subordinate to it.[75]

This debate is closely linked to the various interpretations of Husserl's position on mathematics. It is on the basis of his contention that Husserl takes ontology to be subordinate to phenomenology that van Atten interprets Husserl's overall approach to mathematics as in harmony with Brouwer's. Since formal ontology, which specifies the formal characteristics of objects, is part of the *mathesis universalis*, van Atten concludes that formal mathematics, too, is subordinate to phenomenology, and takes this to entail a revisionist conception of mathematics. According to van Atten, Husserl's position in *Formal and Transcendental Logic* is that formal ontology is subordinate to the logic of truth, and not only to the logic of noncontradiction (van Atten 2010, 47). If, on the other hand, formal ontology is not seen as subject to phenomenological constraints, but as constrained by mathematics (e.g., by a theory of all possible manifolds) and the logic of noncontradiction, Husserl's overall approach to mathematics can be interpreted as in harmony with Hilbert's formalism.

The next two sub-sections explore various conceptions of the relation between phenomenology and ontology, and their relevance to the relation between phenomenology and mathematics.

4.5.1 Being and Time *and the Foundations of Mathematics*

Discussion of the relation between phenomenology and ontology, and of affinities between phenomenology and intuitionism, raises the question of the relation between Heidegger's *Being and Time* and Husserl's philosophy of mathematics. Although *Being and Time* is dedicated to Husserl – indeed, in *Being and Time*, Heidegger asserts that "phenomenology is the science of the Being of entities – ontology" (Heidegger 1962, 61) – its exact relation to Husserl's phenomenology has not ceased to be controversial. Is it

---

[75] D. W. Smith argues for formal ontology's independence of phenomenology (see, e.g., Smith 2003). Drummond critiques Smith's stance, arguing that Husserl takes ontology to be subordinate to phenomenology (Drummond 2009). Smith (2013, chapter 9) responds to this critique.

a phenomenological alternative to Husserl's approach, or does it eventually lead to the rejection of phenomenology? Husserl considered Heidegger his most gifted student, and the one best suited to succeed him at the University of Freiburg, but later came to think that *Being and Time* did not remain on the path of phenomenology as a rigorous science, but represented a philosophical inquiry that, like psychologism, was based on contingent features of human existence (an approach Husserl called "philosophical anthropology"). In this Element, my assumption is that *Being and Time* does not break with phenomenology, but offers a new foundation for it.

Their respective attitudes to mathematics attest to an important difference of opinion between Husserl and Heidegger. For Husserl, mathematics is a paradigm case of knowledge, and closely linked to formal ontology. In *Being and Time*, by contrast, mathematics does not play a privileged role in ontology, which inquires into the meaning of Being. Nevertheless, there are various ways of relating *Being and Time* to phenomenological approaches to the philosophy of mathematics.[76] I focus on two such connections: the relation between *Being and Time* and intuitionism, discussed here, and the relevance of *Being and Time* to the phenomenology of mathematical practice, discussed in Section 4.7.

There are two principal links between *Being and Time* and intuitionism. First, like Brouwer, Heidegger rejects the unqualified validity of the law of excluded middle. Heidegger's position is based on an examination of Husserl's notion of truth as the fulfillment of intention. He argues that this notion entails that there is no justification for separating truth from its givenness to consciousness. Heidegger famously gives the example of Newton's laws, which, before being discovered by Newton, were neither true nor false (Heidegger 1962, 269).

Another link between *Being and Time* and intuitionism concerns the notion of time. According to Heidegger, "time is the transcendental horizon for the question of Being" (Heidegger 1962, 63). This conception of time, in asserting an intimate connection between time and entities – presumably including mathematical entities – calls to mind Brouwer's conception of time. But the two notions of time differ. Heideggerian time is not simply the "falling apart of moments of life" (Brouwer 1975, 127) and their re-unification. More fundamental for Heidegger is finite time, which is oriented toward the inevitable death awaiting us in the future. According to Heidegger, as the ultimate horizon for all our possible activities, death is the basis for all possible meanings of the notion of time. It will be recalled that Becker drew a connection between a central meaning of time discussed in *Being and Time* – historical time – and

---

[76] See Roubach (2008).

choice sequences. In my opinion, Heidegger's notion of finite time can, like his notion of historical time, contribute to philosophical elucidation of the finite perspective underlying intuitionism. This perspective brings to the fore the idea that choice sequences are objects that are never completely determined. As I see it, intuitionism's finite perspective requires a notion of finite time, such as Heidegger's notion of finite time oriented toward our impending death.[77] Dasein, as a living being, can exist only as incomplete, just as choice sequences are inherently incomplete.

Is Heidegger's claim about the temporality of all entities, including mathematical entities, equivalent to Husserl's claim that mathematical objects are not a-temporal, but omni-temporal? It is not, since whereas both Brouwer and Heidegger take temporality to entail incompleteness, Husserl does not. Because Husserl takes the structure of time (e.g., every temporal experience includes retention of the near past and protention toward the future) to be the basis for the relation between mathematics and time, and a full characterization of this structure can be supplied, Husserl's conception of temporality does not entail incompleteness.

4.5.2 Mathematics and the Externalist Interpretation of Phenomenology

Two questions arise regarding the relation between phenomenology and ontology. First, how is phenomenology related to the basic characterizations of entities? Section 4.5 opened with a discussion of that question. Second, what, precisely, is the relation between things and their givenness to consciousness? Can things be separated from their givenness? Husserl speaks of the correlation between things and their givenness to consciousness, meaning that we apprehend things as they are, as the basic tenet of his philosophy (*Crisis*, §48). But this correlation can be understood in different ways. It can be understood as an idealist position that denies that things exist independently of their relation to consciousness, and it can be understood as a realist position according to which consciousness captures things as they are in themselves. The discussion about Platonistic and intuitionistic interpretations of Husserl's stance on the existence of mathematical concepts reflects these opposed understandings of the relation between things and their givenness to consciousness. Platonistic interpretations of Husserl's stance on mathematical concepts are anchored in a realistic interpretation of the thing–givenness relation, whereas intuitionistic interpretations are anchored in an idealist interpretation of that relation.

---

[77] See Posy (2020, 90–6) for a characterization of this intuitionistic finite perspective, and Roubach (2008, 64–9) on the connections between intuitionism and Heidegger's notion of finite time.

The debate over the correct interpretation of phenomenology can be traced back to the initial phase of the reception of Husserl's phenomenology. Since the 1970s, the debate about whether Husserl's position on the objects of intentions is idealistic or realistic has produced opposed interpretations known as the "west coast" and "east coast" approaches. The chief focus of the dispute between these approaches is the relation between an intention's meaning (the noema), and the thing toward which the intention is directed. The west coast approach contends that the noema is distinct from the thing toward which the intention is directed. In support of this position, Husserl's statement that a perceived tree can burn, but the sense of the perception of the tree cannot (*Ideas*, §89) can be adduced. The east coast approach contends that there is no such separation between the noema and that toward which the intention is directed.[78] In addition to its salience vis-à-vis the question of whether Husserl's position on the objects of intentions should be interpreted as idealistic or realistic, the west coast–east coast debate has consequences for how Husserl's "bracketing" (*epoché*) should be understood. Does it also have consequences for the phenomenological understanding of the ontological status of mathematical notions? Are the mathematical notions some sort of external objects? Noemata? Or are they among the formal conditions for thought?

Two interpretations of Husserl's position on the ontological status of the mathematical notions are, I suggest, pertinent to the west coast–east coast debate over the relation between noemata and objects. One such interpretation, the "application of mathematics first" approach mentioned in Section 3.4, takes the application of mathematics to be the starting point for phenomenology of mathematics. It takes the mathematical notions to be extracted from the world, which they are an integral aspect of. As such, it can be deemed an Aristotelian conception of mathematics.[79] The interpretation of Husserl's conception of natural numbers on the basis of Fine's analysis, discussed in Section 3.2.1, exemplifies such an Aristotelian orientation, since it extracts the concept of number from concrete multiplicities. Although several core aspects of Husserl's thought, such as his claim that categorial intuition is founded on sensory intuition, are in line with the Aristotelian approach to mathematics, other aspects, such as his claim that mathematics has a purely formal dimension, are not.[80]

---

[78] For a defense of this position, see Drummond (1990).

[79] Cobb-Stevens (2002) presents an Aristotelian interpretation of Husserl's position in *Logical Investigations*.

[80] An Aristotelian interpretation of Husserl's position can be developed without assuming that ontology is subordinate to phenomenology. D. W. Smith takes Husserl's theory of ideal objects – the justification for which is anchored in the part–whole relation, a central component of Husserl's formal ontology – to be Aristotelian (Smith 2013, 160).

The second interpretation of Husserl's phenomenology of mathematics that is pertinent to the west coast–east coast debate is best understood against the background of the distinction, widely upheld in analytic philosophy, between externalism and internalism. This distinction has been articulated in various ways; here, I adduce Putnam's version. Putnam offers an account of externalism – as he puts it, the position that "meanings just ain't in the head" (Putnam 1975, 227) – in terms of the distinction between intension and extension. Invoking the famous example of the meaning (or intension) of "water," Putnam argues that either the idea that intensions are determined by psychological states, or the idea that intension determines extension, should be given up (Putnam 1975, 222). The rejection of these ideas can be related to the interpretation of Husserlian noemata. The east coast position that noemata are not separate from objects in the world calls to mind the idea that intensions do not determine extensions – that is, the idea that the internal doesn't determine the external.[81]

Externalism can, I contend, be related to Pradelle's interpretation of Husserl's phenomenological approach to mathematical concepts as going beyond Kant's Copernican revolution.[82] On Pradelle's interpretation, the subject mirrors transcendent objects, not the other way around. Borrowing Putnam's formulation, Pradelle's interpretation of Husserl entails rejection of the claim that intensions are determined by psychological states. Pradelle argues that Husserl, instead of taking objects to be determined by the cognizing subject (Kant's Copernican revolution), contends that integral to every domain or type of objects is the mode of its givenness to consciousness, or to put it differently, every domain or type of objects provides clues as to its transcendental investigation (see *CM*, §22). The claim that Husserl goes beyond the Kantian Copernican revolution raises the question of whether Husserl succeeds in avoiding reversion to the pre-Kantian position that the world is exactly as we experience it (naive realism). With respect to mathematical concepts, the question is whether – and if so, how – Husserl's position differs from Platonism. Pradelle's answer is that Husserl rejects the idea that the intuition of mathematical concepts, like the intuition of sensible objects, is some sort of perception. Husserlian intuition in mathematics, Pradelle maintains, is tantamount to demonstration, not only in the sense of providing a proof, but also in the sense of relating new notions to

---

[81] On the relevance of the internalism–externalism distinction to Husserl's thought, see Zahavi (2017, chapter 4).

[82] See Pradelle (2012). Pradelle's focus is not the original Copernican revolution, but the Kantian one. In 1934 Husserl wrote a short text entitled "Overthrow of the Copernican Theory"; it was published in 1940 under a different title. Its core thesis is that our experience of the world, including the experience that the earth does not move, should be the starting point of phenomenology.

already familiar notions. Intuition in mathematics is not some sort of perception, but rather fulfillment on the basis of evidence (i.e., proofs) (Pradelle 2020, 133–4). As Pradelle understands Husserl, evidence evolves over time and is intersubjective. Hence, Pradelle takes Husserl's notion of evidence to be Leibnizian, not Cartesian.

4.6 Phenomenology and the French School of Philosophy of Mathematics

Within nonanalytic twentieth-century philosophy, the philosophy of mathematics is a central focus of just two schools of thought: phenomenology, and the "French epistemological school." The core insight of the latter school is that philosophy of science develops historically together with the development of science. Philosophy of science does not examine science from an external perspective, or attempt to impose, one way or another, any norms on the activity of scientists. Leon Brunschvicg, a founder of this approach, was the first to apply this insight to mathematics, in his 1912 *Les étapes de la philosophie mathématique*, asserting that the philosophy of mathematics develops historically together with the development of mathematics.

The thought of Jean Cavaillès, a member of the French epistemological school and a philosopher of mathematics, played a key role in French philosophy after World War II. Cavaillès explicitly critiques Husserl's position on logic and mathematics. Cavaillès's philosophical approach seeks to maintain the autonomy of mathematical becoming. On his view, the development of mathematics is not constrained by a specific conception of logic (e.g., acceptance or rejection of the law of excluded middle) or account of its origin (e.g., in Euclidean geometry). Cavaillès sees Husserl's phenomenology as undermining this autonomy. He discusses Husserl's position chiefly in *On Logic and the Theory of Science* (Cavaillès 2021), which examines the historical development of logic as a theory of science from Kant onward. Logic as a theory of science not only describes the structure of thought in general, but also plays an active role in acquiring and establishing knowledge.

According to Cavaillès, any attempt to understand logic as the foundation for a theory of science is challenged by two conflicting demands: logic's necessity must be preserved, yet at the same time, logic must be grounded in the subject. These conflicting requirements generate a tension that, after Kant, dissolves into two opposed approaches to logic. Some philosophers, among them Bolzano, Frege, and Carnap, stress logic's objective validity; others, such as Brunschvicg and Brouwer, stress that logic is ultimately grounded in the subject. Each approach must contend with an unresolved problem. Those who stress logic's objective validity must account for the fact that logic nonetheless evolves over time. Given the objective validity of logic, how is the change from one conception of logic to another to be understood? Those who emphasize the relation of logic to a knowing subject must find a way to explain logic's objective validity. A consequence of their position is that the laws of logic are grounded in something external, namely, the knowing subject, hence the validity of these laws is open to question. According to Cavaillès, Husserl tried to reconcile these two conceptions of logic, but did not succeed in devising a unified conception (Cavaillès 2021, 93). The primary impediment, Cavaillès argues, was that in effect, Husserl's approach to logic and consciousness imposed constraints on mathematics. One such constraint was Husserl's characterization of formal systems in terms of the notion of a definite manifold. Cavaillès claims that these constraints, which were unnecessary, interfered with the development of mathematics. Rejecting Husserl's focus on consciousness, he concludes: "It is not a philosophy of consciousness but a philosophy of the concept that can yield a doctrine of science" (Cavaillès 2021, 136). In other words, Cavaillès's critique amounts to the claim that Husserl's philosophy of mathematics is weakly revisionist.

Cavaillès argues that Husserlian phenomenology, in effect, imposes constraints on the development of mathematics, thereby impeding it. Nevertheless, there are, I contend, affinities between Cavaillès's outlook and at least some of Husserl's claims about mathematics. Consider, first, Cavaillès's principle that there should be no external constraints on mathematics, and hence philosophy of mathematics should focus on the actual thought of mathematicians. This is in line with phenomenology's stance. Second, there is some similarity between Cavaillès's historical approach and Husserl's historical conception of mathematics in *The Crisis of European Sciences* and "Origin of Geometry." But this similarity is limited, since Cavaillès claims that in these texts Husserl does not provide an adequate justification for the historical approach (Cavaillès 2021, 135). It could also be argued, in the spirit of Cavaillès's critique, that Husserl's theory of geometry's origin imposes unwarranted constraints on the development of geometry.

Jean-Toussaint Desanti, who studied under Cavaillès, helped diminish the gaps between phenomenology and the Cavaillèsian approach. Desanti characterizes his work as an investigation of the way mathematical objects manifest themselves (Desanti 1968, 289); this is a patently phenomenological characterization. Desanti's main critique of Husserl's phenomenology targets the notions of the transcendental ego and constitutive subject. The rationale underlying this critique, as in the case of Cavaillès's critique, is rejection of external constraints on the development of mathematics. *Logical Investigations*, which does not posit a transcendental ego, might seem to be in harmony with the "no external constraints" requirement. But even this early Husserlian stance cannot be accepted. Vis-à-vis mathematical knowledge, Desanti maintains that there is no basic level with respect to which we have immediate intuitions: there is always a field of meaning in which our intuitions are embedded. The Husserlian requirement of givenness in intuition is, according to Desanti, an unjustified constraint on mathematics.

Maurice Caveing, Desanti's disciple, argues that formulation of an adequate philosophy of mathematics requires us to relinquish three Husserlian ideas: the idea that a fundamental epistemological grounding of all knowledge can be provided; the idea that the transcendental ego is the source of mathematics; and the idea that a complete description of mathematics can be given (Caveing 2004, 14).

Two aspects of Pradelle's interpretation of Husserl's position, discussed in Section 4.5.2, further contribute to closing the gaps between Husserl's phenomenology and the French epistemological school. Pradelle's characterization of phenomenology as going beyond Kant's Copernican revolution, inasmuch as it sees the mode of givenness of the mathematical notions as inherent in them, can be construed as compatible with Desanti's critique of the transcendental ego and Cavaillès's rejection of external constraints on mathematics. And Pradelle's interpretation of Husserlian intuition in terms of proofs rather than as a species of perception renders it acceptable from the perspective of the French epistemological school, since it allows for the historical development of intuition through the work of mathematicians.

4.7 Phenomenology and the Philosophy of Mathematical Practice

In recent decades analytic philosophy of mathematics has witnessed the emergence of an approach that regards the practice of mathematicians as the basis for philosophical reflection on mathematics.[83] The underlying idea is that philosophy of mathematics should track the practice of mathematicians, rather than impose philosophical preconceptions on mathematics. This new approach has, in my opinion, a pronounced affinity with the French epistemological school, and as I will argue, phenomenology can be interpreted as a philosophy of mathematical practice.

The affinity between phenomenology and the philosophy of mathematical practice is not surprising. It can be discerned in Husserl's emphasis on attending to the perspective of mathematicians, his division of labor between philosophy

---

[83] See, for example, Mancosu (2008). For a recent overview of the philosophy of mathematical practice, see Carter (2019).

and mathematics, his assertion that he accepts the existence of numbers because of the way mathematicians speak (Husserl 1975, 39), and the fact that phenomenology's declared goal is to describe things as they are, without preconceptions.[84]

As further support for the suggested affinity, the philosophical work of Gian-Carlo Rota, a leading twentieth-century mathematician, can be adduced. Rota approaches the philosophy of mathematics from a phenomenological perspective, yet can also be interpreted as endorsing a philosophy of mathematical practice. Rota undermines the claim that formal logic is the unique ground for acceptance of mathematical theorems. He distinguishes between mathematical proofs and mathematical truth (Rota 1990, 267). Having a proof is not always sufficient for a theorem to be accepted as true, and truth is subordinate to evidence, evidence being understood by Rota as the experience of grasping that a statement must be true (Rota 1997b, 169).[85] Only when mathematicians attain evidence for it are mathematicians satisfied that a theorem is true. If the proofs that exist at a certain point are not deemed sufficiently evident, mathematicians will go on searching for a proof that meets this standard (Rota 1991, 135). The centrality of evidence to Husserl's understanding of truth can be seen as reflecting the practice of mathematicians as Rota describes it. Rota's characterization of his philosophy of mathematics as descriptive (Rota 1997a, 184) is also in line with both Husserl's phenomenology, and the philosophy of mathematical practice.[86]

Rota's approach to the philosophy of mathematics mainly engages with Husserl's phenomenology, but *Being and Time* also had a significant influence on it. The central idea Rota adopts from *Being and Time* is that our relation to the world is primarily one of coping, and we relate to objects chiefly as tools. Rota's understanding of the mathematical concepts is much like Heidegger's understanding of objects as ready-to-hand. *Being and Time*'s critique of the primacy given to subject–predicate judgments by traditional philosophy is also invoked by Rota in critiquing contemporary analytic philosophy of mathematics, in which formal logic plays a key role.[87]

---

[84] Leng (2002) discusses affinities between phenomenology and the philosophy of mathematical practice.

[85] Rota's approach to mathematical knowledge, in giving evidence primacy over truth, might seem somewhat similar to Brouwer's conception, but it is difficult to make the comparison, since Rota's notion of evidence is Cartesian (i.e., related to certainty), whereas Brouwer's is general.

[86] From the perspective of the philosophy of mathematical practice, Husserl's characterizations of mathematics – for example, the characterization of an axiomatic system as definite – are not constraints on mathematics, as some of the French epistemologists claim, but rather describe the practices of mathematicians in various historical periods. This is, I contend, Hartimo's view.

[87] See Rota (1997b, 188–91).

Rota denies that clear distinctions can be drawn between ideal and real objects and between mathematics and the empirical sciences (Rota 1997b, 167). He interprets Husserl as promoting "non-normative," that is, factual, descriptions of mathematical objects (164). I would argue that this position misconstrues the place of normativity in Husserl's phenomenological approach to mathematics. Husserl thinks of mathematical objects as ideal, this ideality being related to the normative role of these objects. Moreover, interpreting mathematical objects as ideal is in keeping with the orientation of the philosophy of mathematical practice. Husserl's notion of intention, which entails a distinction between the ideal and the real, can make an important contribution to the philosophy of mathematical practice, since it enables us to describe the mathematician's activities without having to relinquish the ideality of mathematical objects. The claim that our focus should be mathematical practice does not entail that mathematical practice has no normative or ideal dimension: according to Husserl, all human activities that have an intentional element involve norms, hence mathematics involves norms.

Hartimo (2021) has suggested another interpretation of Husserl's thought on mathematics that underscores its affinities with the philosophy of mathematical practice. According to Hartimo, Husserl gives priority to the discipline of mathematics itself, and the role of the phenomenological investigation of mathematics is to engage in reflection (*Besinnung*) on mathematics and examine whether it fulfills its aims. Hartimo rejects the idea that Husserl's approach to mathematics exemplifies one particular philosophical outlook, arguing that it is "a combination of constructivism, various kinds of structuralism, and Platonism" (Hartimo 2021, 173). This pluralistic interpretation of Husserl's philosophy of mathematics is linked to mathematical practice. As Mancosu (2008, 7) argues, the foundational projects of Frege, Hilbert, and Brouwer all have roots in the thought and practice of nineteenth-century mathematics. At the time these philosophical positions were conceived, the practice of mathematics was nonexclusive, and had logicistic, formalistic, and intuitionistic features. Husserl's philosophy of mathematics can be seen as adopting this nineteenth-century attitude.

Interpreting Husserl's position as aligned with the spirit of the philosophy of mathematical practice suggests answers to the question of how phenomenology can be perceived as having affinities with three conflicting positions in the philosophy of mathematics – Platonism, formalism, and intuitionism. Hartimo's claim that phenomenology's conception of mathematics is pluralistic is one such answer. Another way to meet the challenge of the conflicting interpretations of Husserl's approach to mathematics is to argue, in line with Pradelle's view, that the different modes of givenness of different species of mathematical entities might fit locally with a Platonistic, formalistic, or constructivist approach to mathematics. We saw this in Section 3.3.2 vis-à-vis geometry.

Both Hartimo and Pradelle base their interpretations on a close reading of *Formal and Transcendental Logic*, in which the different layers of logic can be interpreted as supporting such pluralism. Confirmation of the idea that phenomenology is not committed to a single philosophical orientation vis-à-vis mathematics can also be found in an unpublished text from 1932, in which Husserl says that we ought not adopt any specific philosophical approach to mathematics, such as Hilbert's or Brouwer's, since we *know* what mathematics is.[88] The idea that phenomenology is not committed to any one philosophical outlook can also be linked to a more general reading of Husserl's phenomenology as metaphysically neutral. On this reading, phenomenology makes no ontological commitment to the existence of essences in general and mathematical notions in particular.[89] We must, however, keep in mind that formalism can in any event be taken to be metaphysically neutral. The hypothesis that phenomenology is metaphysically neutral is thus relevant primarily to attempts to argue for a Platonistic interpretation of Husserl's stance, and to some extent also to arguments for constructivist interpretations.

Another strategy for resolving the problem of the conflicting interpretations of Husserl's approach to mathematics is to argue that Husserl's phenomenological approach to mathematics is not pluralistic, but a unique combination of formalism, Platonism, and intuitionism. Mathematics has formalistic aspects (e.g., the notion of a formal multiplicity), but it is not solely formalistic, since fulfillment of intentions (intuition) also has a place in mathematics. Intuition should not be understood as attesting to either the independent existence of the mathematical notions (Platonism) or some sort of process of constructing these notions, but rather, as demonstrating a combination of Platonism and constructivism. Tieszen refers to this position as "constituted realism." It is akin to Derrida's interpretation of the stance Husserl upholds in "Origin of Geometry," and can also be seen as reflecting mathematical practice.

But whether we adopt one of these resolutions of the conflicting interpretations of the phenomenology of mathematics, or opt to uphold one of the interpretations, the questions initially raised by Husserl – questions that motivate any phenomenological examination of mathematics – must be acknowledged. How is mathematics given to consciousness, or to put it differently, what are the experiences that are inextricably connected to mathematics? How should we understand the distinction between meaning and intuition in mathematics? And what counts as evidence in mathematics?

---

[88] See van Atten (2007, 64–5) for an English translation of this text.

[89] On the metaphysical neutrality of Husserlian phenomenology, see Zahavi (2017, chapter 2). Thomasson (2017) interprets Husserl's position as not committed to the metaphysical existence of essences.

Abbreviations

Britannica   Husserl, E. (1981). "Phenomenology," Edmund Husserl's Article for the *Encyclopedia Britannica* (1927). Translated by R. E. Palmer. In P. McCormick & F. Elliston (eds.), *Husserl: Shorter Works*. Notre Dame, IN: University of Notre Dame Press, 21–35.

CM   Husserl, E. (1960). *Cartesian Meditations. An Introduction to Phenomenology*. Translated by D. Cairns. The Hague: Martinus Nijhoff.

Crisis   Husserl, E. (1970). *The Crisis of European Sciences and Transcendental Phenomenology: An Introduction to Phenomenological Philosophy*. Translated by D. Carr. Evanston, IL: Northwestern University Press.

EJ   Husserl, E. (1973). *Experience and Judgment. Investigations in a Genealogy of Logic*. Translated by J. S. Churchill & K. Ameriks. Evanston, IL: Northwestern University Press.

FTL   Husserl, E. (1969). *Formal and Transcendental Logic*. Translated by D. Cairns. The Hague: Martinus Nijhoff.

Ideas   Husserl, E. (1998). *Ideas Pertaining to a Pure Phenomenology and to a Phenomenological Philosophy. First Book: General Introduction to a Pure Phenomenology*. Translated by F. Kersten. Dordrecht: Kluwer.

LI   Husserl, E. (2001). *Logical Investigations*. Translated by J. N. Findlay. Edited by D. Moran. New York: Routledge.

PA   Husserl, E. (2003). *Philosophy of Arithmetic: Psychological and Logical Investigations with Supplementary Texts from 1887–1901*. Translated by D. Willard. Dordrecht: Springer. https://doi.org/10.1007/978-94-010-0060-4.

References

van Atten, M. (2007). *Brouwer Meets Husserl: On the Phenomenology of Choice Sequences*. Dordrecht: Springer. https://doi.org/10.1007/978-1-4020-5087-9.

van Atten, M. (2010). Construction and Constitution in Mathematics. *New Yearbook for Phenomenology and Phenomenological Philosophy*, 10: 43–90.

van Atten, M., van Dalen, D., and Tieszen, R. (2002). Brouwer and Weyl: The Phenomenology and Mathematics of the Intuitive Continuum. *Philosophia Mathematica*, 10: 203–26. https://doi.org/10.1093/philmat/10.2.203.

van Atten, M., and Kennedy, J. (2003). On the Philosophical Development of Kurt Gödel. *Bulletin of Symbolic Logic*, 9: 425–76.

Bachelard, S. (1968). *A Study of Husserl's Formal and Transcendental Logic*. Translated by L. E. Embree. Evanston, IL: Northwestern University Press.

Becker, O. (1923). Beiträge zur Phänomenologischen Begründung der Geometrie und ihrer physikalischen Anwendungen. *Jahrbuch für Philosophie und phänomenologische Forschung*, 6: 385–560.

Becker, O. (1927). *Mathematische Existenz. Untersuchungen zur Logik und Ontologie mathematischer Phänomene*. Halle: Max Niemeyer.

Benacerraf, P. (1973). Mathematical Truth. *Journal of Philosophy*, 70: 661–79.

Brouwer, L. E. J. (1975). Intuitionism and Formalism. *Bulletin of the American Mathematical Society*, 20: 81–96. https://doi.org/10.1090/S0002-9904-1913-02440-6.

Cantor, G. (1915). *Contributions to the Founding of the Theory of Transfinite Numbers*. Translated by P. Jourdain. New York: Dover.

Carter, J. (2019). Philosophy of Mathematical Practice: Motivations, Themes and Prospects. *Philosophia Mathematica*, 27: 1–32. https://doi.org/10.1093/philmat/nkz002.

Cavaillès, J. (2021). *On Logic and the Theory of Science*. Translated by R. Mackay and K. Peden. Falmouth: Urbanomic Media.

Caveing, M. (2004). *Le problème des objets dans la pensée mathématique*. Paris: Vrin.

Centrone, S. (2010). *Logic and Philosophy of Mathematics in the Early Husserl*. Dordrecht: Springer.

Cobb-Stevens, R. (2002). Aristotelian Themes in Husserl's *Logical Investigations*. In D. Zahavi and F. Stjernfelt (eds.), *One Hundred Years of Phenomenology: Husserl's Logical Investigations Revisited*. Dordrecht: Kluwer, 79–92. https://doi.org/10.1007/978-94-017-0093-1_6.

Da Silva, J. J. (2000). Husserl's Two Notions of Completeness: Husserl and Hilbert on Completeness and Imaginary Elements in Mathematics. *Synthese*, 125: 417–38.

Da Silva, J. J. (2013). How Sets Came to Be: The Concept of Set from a Phenomenological Perspective. *New Yearbook for Phenomenology and Phenomenological Philosophy*, 13: 84–100.

Da Silva, J. J. (2016). Husserl and Hilbert on Completeness, Still. *Synthese*, 193: 1925–47. https://doi.org/10.1007/s11229-015-0821-2.

Derrida, J. (1989). *Edmund Husserl's Origin of Geometry: An Introduction*. Translated by J. P. Leavey Jr. Lincoln, NB: University of Nebraska Press.

Desanti, J.-T. (1968). *Les idéalités mathématiques*. Paris: Seuil.

Descartes, R. (1985). Rules for the Direction of the Mind. Translated by D. Murdoch. In J. Cottingham, R. Stoothoff, and D. Murdoch (eds.), *The Philosophical Writings of Descartes*, vol. 1. Cambridge: Cambridge University Press, 7–78. https://doi.org/10.1017/CBO9780511805042.004.

Drummond, J. J. (1990). *Husserlian Intentionality and Non-Foundational Realism*. Dordrecht: Kluwer.

Drummond, J. J. (2009). Phénoménologie et ontologie. Translated by G. Fréchette. *Philosophiques*, 36: 593–607. https://doi.org/10.7202/039488ar.

Fine, K. (1998). Cantorian Abstraction: A Reconstruction and Defense. *Journal of Philosophy*, 95 (12): 599–634. https://doi.org/10.2307/2564641.

Føllesdal, D. (1994). Husserl and Frege: A Contribution to Elucidating the Origins of Phenomenological Philosophy. Translated by C. O. Hill. In L. Haaparanta (ed.), *Mind, Meaning, and Mathematics: Essays on the Philosophical Views of Husserl and Frege*. Dordrecht: Kluwer, 3–47. https://doi.org/10.1007/978-94-015-8334-3_1.

Frege, G. (1972). Review of Husserl's *Philosophy of Arithmetic*. Translated by E. W. Kluge. *Mind*, 81: 321–37.

Frege, G. (1980a). Begriffsschrift, a Formula Language, Modeled upon That of Arithmetic, for Pure Thought. In J. van Heijenoort (ed.), *Frege and Gödel: Two Fundamental Texts in Mathematical Logic*. Cambridge, MA: Harvard University Press, 1–82.

Frege, G. (1980b). *The Foundations of Arithmetic: A Logico-Mathematical Inquiry into the Concept of Number*. Translated by J. L. Austin. Evanston, IL: Northwestern University Press.

Gödel, K. (1961). The Modern Development of the Foundations of Mathematics in the Light of Philosophy. In K. Gödel, S. Feferman, J. W. Dawson Jr., W. Goldfarb, C. Parsons, and R. N. Solovay (eds.), *Collected Works*, vol. 3 (1995). Oxford: Oxford University Press, 374–87.

Gödel, K. (1964). What is Cantor's Continuum Problem? In K. Gödel, S. Feferman, J. W. Dawson Jr., W. Goldfarb, C. Parsons, and R. N. Solovay (eds.), *Collected Works*, vol. 2 (1990). Oxford: Oxford University Press, 254–70.

Hartimo, M. (2007). Towards Completeness: Husserl on Theories of Manifolds 1890–1901. *Synthese*, 156: 281–310. https://doi.org/10.1007/s11229-006-0008-y.

Hartimo, M. (2018). Husserl on Completeness, Definitely. *Synthese*, 195: 1509–27. https://doi.org/10.1007/s11229-016-1278-7.

Hartimo, M. (2021). *Husserl and Mathematics*. Cambridge: Cambridge University Press. https://doi.org/10.1017/9781108990905.

Hauser, K. (2006). Gödel's Program Revisited Part I: The Turn to Phenomenology. *Bulletin of Symbolic Logic*, 12: 529–90. https://doi.org/10.2178/bsl/1164056807.

Heidegger, M. (1962). *Being and Time*. Translated by J. Macquarrie and E. Robinson. Oxford: Blackwell.

Hilbert, D. (1964). On the Infinite. In P. Benacerraf and H. Putnam (eds.), *Philosophy of Mathematics: Selected Readings*. Englewood Cliffs, NJ: Prentice-Hall, 134–51.

Hilbert, D. (1996). On the Concept of Number. Translated by W. Ewald. In W. Ewald (ed.), *From Kant to Hilbert*, vol. 2. Oxford: Oxford University Press, 1092–5.

Hill, C. O. (2000). Abstraction and Idealization in Georg Cantor and Edmund Husserl Prior to 1895. In C. O. Hill and G. E. Rosado Haddock (eds.), *Husserl or Frege: Meaning, Objectivity, and Mathematics*. Chicago: Open Court, 109–36.

Hill, C. O. (2010). Husserl on Axiomatization and Arithmetic. In M. Hartimo (ed.), *Phenomenology and Mathematics*. Dordrecht: Springer, 47–71. https://doi.org/10.1007/978-90-481-3729-9_3.

Hintikka, J. (2003). The Notion of Intuition in Husserl. *Revue internationale de philosophie*, 224: 169–91.

Hopkins, B. C. (2005). Klein and Derrida on the Historicity of Meaning and the Meaning of Historicity in Husserl's *Crisis*-Texts. *Journal of the British Society for Phenomenology*, 36: 179–87. https://doi.org/10.1080/00071773.2005.11006541.

Hopkins, B. C. (2011). *The Origin of the Logic of Symbolic Mathematics: Edmund Husserl and Jacob Klein*. Bloomington, IN: Indiana University Press.

Husserl, E. (1956). *Erste Philosophie. Erster Teil: Kritische Ideengeschichte*. The Hague: Martinus Nijhoff.

Husserl, E. (1975). *Introduction to the Logical Investigations: A Draft of a Preface to the Logical Investigations.* Edited by E. Fink. Translated by P. J. Bossert and C. H. Curtis. The Hague: Martinus Nijhoff.

Husserl, E. (1980). *Ideas III: Phenomenology and the Foundations of the Sciences.* Translated by T. Klein and W. Pohl. The Hague: Martinus Nijhoff.

Husserl, E. (1983). *Studien zur Arithmetik und Geometrie. Texte aus dem Nachlass (1886–1901).* Edited by I. Strohmeyer. The Hague: Martinus Nijhoff.

Husserl, E. (1991). *On the Phenomenology of the Consciousness of Internal Time (1893–1917).* Translated by J. B. Brough. Dordrecht: Kluwer.

Husserl, E. (1994). *Early Writings in the Philosophy of Logic and Mathematics.* Translated by D. Willard. Dordrecht: Kluwer.

Husserl, E. (1997). *Thing and Space: Lectures from 1907.* Translated by R. Rojcewicz. Dordrecht: Springer.

Ierna, C. (2017). The Brentanist Philosophy of Mathematics in Edmund Husserl's Early Works. In S. Centrone (ed.), *Essays on Husserl's Logic and Philosophy of Mathematics.* Dordrecht: Springer, 147–68. https://doi.org/10.1007/978-94-024-1132-4_7.

Ierna, C., and Lohmar, D. (2016). Husserl's Manuscript A I 35. In G. E. Rosado Haddock (ed.), *Husserl and Analytic Philosophy.* Berlin: De Gruyter, 289–320. https://doi.org/10.1515/9783110497373-011.

Kant, E. (1998). *Critique of Pure Reason.* Translated by P. Guyer and A. W. Wood. Cambridge: Cambridge University Press.

Klein, J. (1940). Phenomenology and the History of Science. In M. Farber (ed.), *Philosophical Essays in Memory of Edmund Husserl.* Cambridge, MA: Harvard University Press, 143–63. https://doi.org/10.4159/harvard.97806743 33512.c8.

Leng, M. (2002). Phenomenology and Mathematical Practice. *Philosophia Mathematica*, 10: 3–25.

Linnebo, Ø. (2018). Platonism in the Philosophy of Mathematics. In E. N. Zalta (ed.), *Stanford Encyclopedia of Philosophy* (Spring 2018 Edition). https://plato.stanford.edu/archives/spr2018/entries/platonism-mathematics/.

Lohmar, D. (1990). Wo lag der Fehler der kategorialen Repräsentation? Zu Sinn und Reichweite einer Selbstkritik Husserls. *Husserl Studies*, 7: 179–97. https://doi.org/10.1007/BF00347584.

Lohmar, D. (1993). On the Relation of Mathematical Objects to Time: Are Mathematical Objects Timeless, Overtemporal or Omnitemporal? *Journal of the Indian Council of Philosophical Research*, 10: 73–87.

Lohmar, D. (2000). *Edmund Husserls "Formale und transzendentale Logik."* Darmstadt: Wissenschaftliche Buchgesellschaft.

Lohmar, D. (2004). The Transition of the Principle of Excluded Middle from a Principle of Logic to an Axiom: Husserl's Hesitant Revisionism in Logic. *New Yearbook for Phenomenology and Phenomenological Philosophy*, 4: 53–68.

Maddy, P. (1980). Perception and Mathematical Intuition. *Philosophical Review*, 89: 163–96. https://doi.org/10.2307/2184647.

Mahnke, D. (1917). *Eine Neue Monadologie*. Berlin: Reuther & Reichard.

Mahnke, D. (1966). From Hilbert to Husserl: First Introduction to Phenomenology, Especially that of Formal Mathematics. Translated by D. L. Boyer. *Studies in History and Philosophy of Science*, 8: 75–84. https://doi.org/10.1016/0039-3681(77)90020-6.

Majer, U. (1997). Husserl and Hilbert on Completeness: A Neglected Chapter in Early Twentieth Century Foundations of Mathematics. *Synthese*, 110: 37–56.

Mancosu, P. (ed.). (2008). *The Philosophy of Mathematical Practice*. Oxford: Oxford University Press. https://doi.org/10.1093/acprof:oso/9780199296453.001.0001.

Mancosu, P. (2018). Explanation in Mathematics. In E. N. Zalta (ed.), *Stanford Encyclopedia of Philosophy* (Summer 2018 Edition). https://plato.stanford.edu/archives/sum2018/entries/mathematics-explanation/.

Mancosu, P., and Ryckman, T. (2002). Mathematics and Phenomenology: The Correspondence between O. Becker and H. Weyl. *Philosophia Mathematica*, 10: 130–202.

Miller, J. P. (1982). *Numbers in Presence and Absence: A Study of Husserl's Philosophy of Mathematics*. The Hague: Martinus Nijhoff. https://doi.org/10.1007/978-94-009-7624-5.

Mohanty, J. N. (1977). Husserl and Frege: A New Look at Their Relationship. In J. N. Mohanty (ed.), *Readings on Edmund Husserl's Logical Investigations*. The Hague: Martinus Nijhoff, 22–32. https://doi.org/10.1007/978-94-010-1055-9_3.

Mohanty, J. N. (1991). Husserl's Formalism. In T. M. Seebohm, D. Føllesdal, and J. N. Mohanty (eds.), *Phenomenology and the Formal Sciences*. Dordrecht: Kluwer, 93–105. https://doi.org/10.1007/978-94-011-2580-2_7.

Nenon, T. (1997). Two Models of Foundation in the *Logical Investigations*. In B. C. Hopkins (ed.), *Husserl in Contemporary Context*. Dordrecht: Kluwer, 97–114. https://doi.org/10.1007/978-94-017-1804-2_6.

Parsons, C. (1980). Mathematical Intuition. *Proceedings of the Aristotelian Society*, 80: 145–68. https://doi.org/10.1093/aristotelian/80.1.145.

Parsons, C. (2012). Husserl and the Linguistic Turn. In C. Parsons, *From Kant to Husserl: Selected Essays*. Cambridge, MA: Harvard University Press, 190–214. https://doi.org/10.4159/harvard.9780674065420.c13.

Posy, C. J. (2020). *Mathematical Intuitionism*. Cambridge: Cambridge University Press. https://doi.org/10.1017/9781108674485.

Pradelle, D. (2000). *L'archéologie du monde. Constitution de l'espace, idéalisme et intuitionnisme chez Husserl*. Dordrecht: Springer. https://doi.org/10.1007/978-94-024-1586-5.

Pradelle, D. (2012). *Par-delà la révolution copernicienne. Sujet transcendental et facultés chez Kant et Husserl*. Paris: Presses universitaires de France.

Pradelle, D. (2020). *Intuition et idéalités. Phénoménologie des objets mathématiques*. Paris: Presses universitaires de France.

Putnam, H. (1975). The Meaning of "Meaning." In H. Putnam (ed.), *Mind, Language and Reality: Philosophical Papers*, vol. 2. Cambridge: Cambridge University Press, 215–71. https://doi.org/10.1017/CBO9780511625251.

Reck, E. H. (2013). Frege, Dedekind, and the Origins of Logicism. *History and Philosophy of Logic*, 34: 242–65. https://doi.org/10.1080/01445340.2013.806397.

Reid, C. (1970). *Hilbert*. Berlin: Springer. https://doi.org/10.1007/978-3-662-28615-9.

Reinach, A. (1989). Über den Begriff der Zahl. In K. Schuhmann and B. Smith (eds.), *Sämtliche Werke*, vol. 1. Munich: Philosophia Verlag, 515–29.

Rosado Haddock, G. E. (2010). Platonism, Phenomenology, and Interderivability. In M. Hartimo (ed.), *Phenomenology and Mathematics*. Dordrecht: Springer, 23–46. https://doi.org/10.1007/978-90-481-3729-9_2.

Rota, G.-C. (1989). *Fundierung* as a Logical Concept. *The Monist*, 72: 70–7. https://doi.org/10.5840/monist19897218.

Rota, G.-C. (1990). Mathematics and Philosophy: The Story of a Misunderstanding. *Review of Metaphysics*, 44: 259–71.

Rota, G.-C. (1991). Mathematics and the Task of Phenomenology. In T. M. Seebohm, D. Føllesdal, and J. N. Mohanty (eds.), *Phenomenology and the Formal Sciences*. Dordrecht: Kluwer, 133–8. https://doi.org/10.1007/978-94-011-2580-2_9.

Rota, G.-C. (1997a). The Phenomenology of Mathematical Proof. *Synthese*, 111: 183–96.

Rota, G.-C. (1997b). *Indiscrete Thoughts*. Edited by F. Palombi. Boston: Birkhäuser. https://doi.org/10.1007/978-0-8176-4781-0.

Roubach, M. (2008). *Being and Number in Heidegger's Thought*. London: Continuum. https://doi.org/10.5040/9781472546166.

Roubach, M. (2021). Numbers as Ideal Species: Husserlian and Contemporary Perspectives. *New Yearbook for Phenomenology and Phenomenological Philosophy*, 18: 537–45.

Roubach, M. (2022). *Mathesis Universalis* and Husserl's Phenomenology. *Axiomathes*, 32: 627–37. https://doi.org/10.1007/s10516-021-09544-9.

Smith, D. W. (2003). "Pure" Logic, Ontology, and Phenomenology. *Revue internationale de philosophie*, 57 (224/2): 133–56.

Smith, D. W. (2013). *Husserl*, 2nd ed. London: Routledge. https://doi.org/10.4324/9780203742952.

Spiegelberg, H. (ed.). (1971). From Husserl to Heidegger: Excerpts from a 1928 Freiburg Diary by W. R. Boyce Gibson. *Journal of the British Society for Phenomenology*, 2: 58–83. https://doi.org/10.1080/00071773.1971.11006166.

Steiner, M. (1978). Mathematical Explanation. *Philosophical Studies*, 34: 135–51. https://doi.org/10.1007/BF00354494.

Thomasson, A. (2017). Husserl on Essences: A Reconstruction and Rehabilitation. *Grazer Philosophische Studien*, 94: 436–59. https://doi.org/10.1163/18756735-09403008.

Tieszen, R. (1989). *Mathematical Intuition: Phenomenology and Mathematical Knowledge*. Dordrecht: Kluwer. https://doi.org/10.1007/978-94-009-2293-8.

Tieszen, R. (2005). Free Variation and the Intuition of Geometric Essences: Some Reflections on Phenomenology and Modern Geometry. *Philosophy and Phenomenological Research*, 70: 153–73. https://doi.org/10.1111/j.1933-1592.2005.tb00509.x.

Tieszen, R. (2010). Mathematical Realism and Transcendental Phenomenological Idealism. In M. Hartimo (ed.), *Phenomenology and Mathematics*. Dordrecht: Springer, 1–22. https://doi.org/10.1007/978-90-481-3729-9_1.

Tieszen, R. (2011). *After Gödel: Platonism and Rationalism in Mathematics and Logic*. Oxford: Oxford University Press. https://doi.org/10.1093/acprof:oso/9780199606207.001.0001.

Tieszen, R. (2012). Monads and Mathematics: Gödel and Husserl. *Axiomathes*, 22: 31–52. https://doi.org/10.1007/s10516-011-9162-z.

Tragesser, R. S. (1984). *Husserl and Realism in Logic and Mathematics*. Cambridge: Cambridge University Press.

Tragesser, R. S. (1989). Sense Perceptual Intuition, Mathematical Existence, and Logical Imagination. *Philosophia Mathematica*, 2: 154–94. https://doi.org/10.1093/philmat/s2-4.2.154.

Webb, J. (2017). Paradox, Crisis, and Harmony in Phenomenology. In S. Centrone (ed.), *Essays on Husserl's Logic and Philosophy of Mathematics*. Dordrecht: Springer, 353–408. https://doi.org/10.1007/978-94-024-1132-4_14.

Weyl, H. (1928). Diskussionsbemerkungen zu dem zweiten Hilbertschen Vortrag über die Grundlagen der Mathematik. *Abhandlungen aus dem mathematischen Seminar der Hamburgischen Universität*, 6: 86–8. https://doi.org/10.1007/978-3-663-16102-8_2.

Weyl, H. (1949). *Philosophy of Mathematics and Natural Science*. Translated by O. Helmer. Princeton, NJ: Princeton University Press.

Weyl, H. (1987). *The Continuum: A Critical Examination of the Foundation of Analysis*. Translated by S. Pollard and T. Bole. Kirksville, MO: Thomas Jefferson University Press.

Weyl, H. (1998). On the New Foundational Crisis of Mathematics. Translated by B. Müller. In P. Mancosu (ed.), *From Brouwer to Hilbert: The Debate on the Foundations of Mathematics in the 1920s*. Oxford: Oxford University Press, 86–122.

Willard, D. (1980). Husserl on a Logic That Failed. *Philosophical Review*, 89: 46–64. https://doi.org/10.2307/2184863.

Zahavi, D. (2017). *Husserl's Legacy: Phenomenology, Metaphysics, and Transcendental Philosophy*. Oxford: Oxford University Press. https://doi.org/10.1093/oso/9780199684830.001.0001.

Acknowledgments

Work on this book has been generously supported by the Israel Science Foundation, grant number 1277/19.

I would like to thank Carl Posy for his wise counsel, particularly with regard to the book's structure, Nessa Olshansky-Ashtar for outstanding editorial assistance, and my family for their love and support.

Cambridge Elements ⬳

The Philosophy of Mathematics

Penelope Rush
University of Tasmania

From the time Penny Rush completed her thesis in the philosophy of mathematics (2005), she has worked continuously on themes around the realism/anti-realism divide and the nature of mathematics. Her edited collection *The Metaphysics of Logic* (Cambridge University Press, 2014), and forthcoming essay 'Metaphysical Optimism' (*Philosophy Supplement*), highlight a particular interest in the idea of reality itself and curiosity and respect as important philosophical methodologies.

Stewart Shapiro
The Ohio State University

Stewart Shapiro is the O'Donnell Professor of Philosophy at The Ohio State University, a Distinguished Visiting Professor at the University of Connecticut, and a Professorial Fellow at the University of Oslo. His major works include *Foundations without Foundationalism* (1991), *Philosophy of Mathematics: Structure and Ontology* (1997), *Vagueness in Context* (2006), and *Varieties of Logic* (2014). He has taught courses in logic, philosophy of mathematics, metaphysics, epistemology, philosophy of religion, Jewish philosophy, social and political philosophy, and medical ethics.

About the Series

This Cambridge Elements series provides an extensive overview of the philosophy of mathematics in its many and varied forms. Distinguished authors will provide an up-to-date summary of the results of current research in their fields and give their own take on what they believe are the most significant debates influencing research, drawing original conclusions.

Cambridge Elements ☰

The Philosophy of Mathematics

Elements in the Series

A full series listing is available at: www.cambridge.org/EPM